Charles V. Riley

Social insects from psychical and evolutional points of view

Charles V. Riley

Social insects from psychical and evolutional points of view

ISBN/EAN: 9783741105647

Manufactured in Europe, USA, Canada, Australia, Japa

Cover: Foto ©berggeist007 / pixelio.de

Manufactured and distributed by brebook publishing software
(www.brebook.com)

Charles V. Riley

Social insects from psychical and evolutional points of view

VOL. IX, PP. 1-74　　　　　　　　　　APRIL, 1894

PROCEEDINGS

OF THE

BIOLOGICAL SOCIETY OF WASHINGTON

SOCIAL INSECTS FROM PSYCHICAL AND EVOLU-
TIONAL POINTS OF VIEW.*

BY C. V. RILEY, PH. D.

PRELUDE.

FRIENDS AND FELLOW-MEMBERS:

Custom has ordained that the president of the Biological
Society deliver an annual address, and that the public be invited
to listen thereto. This custom, likewise followed by some of our
sister societies, has certain advantages, but also certain disadvantages. Instead of appealing to members only, or treating, in
special and technical way, some subject that intimately concerns
them, the speaker finds it incumbent upon him to popularize his
subject, and to endeavor to interest alike those who are and those
who are not familiar with the science of biology in any of its
special branches. It will be my endeavor to accomplish this dual
task to-night by omitting the reading of the more technical and
detailed portions of this paper, which, though in one sense the
most important, may well be printed in smaller type, as a series
of notes.

My predecessors have generally dealt with the subjects upon
which they were working as specialists, or upon which they were

---

*Annual address of the President of the Society, delivered in the hall of
Columbian University, January 29, 1894. The address was illustrated with
stereopticon views, only a few of which are here reproduced.

known to be authorities. In following this precedent, I am not
unmindful of the fact that the science of entomology in its
more abstruse and technical phases, however fascinating to the
specialist, attracts but little public attention, and that, from
among the myriad forms of life which the entomologist includes
within the scope of his study, there are comparatively few which
interest the intelligent masses or even the general biologist.
Among these few are the social insects, and it is my purpose to
treat of them to-night and see what light we may draw from
them on some of the great questions which now agitate naturalists. By combining the recorded observations and views of others
with some that are original and unpublished, I may, perhaps,
hope to interest all of you.

Before entering on this main topic, however, it has seemed to
me advisable, in view of the character of the audience, to say
something of our society and what it undertakes to do. Biology
is a word of the century, and was first employed by Lamarck
(1801) as a term under which the phenomena of organic nature
could be considered; and by Treviranus (1802) to express the
science that treats of the philosophy of living nature. Syste-
matic zoology and botany have but incidental bearing on biology;
they relate to the framework, the structure, and not to life itself.
Not that I undervalue taxonomy in this connection, for, indeed,
its value is self-evident; but modern biologists are very generally
divided into two camps, viz., those who investigate the different
parts and structures of the organism, or who study the processes
of growth, and those who study more particularly that phase of
the subject which Hæckel called œcology. In the process of
differentiation the term is now, perhaps, more correctly applied
to the study of the development of the type in the past, and of
the individual in the present—not by themselves only, but in
their relations to all other forms of life. In other words, it involves the interactions and interrelations of organisms, and deals
fundamentally with psychical even more than with structural
phenomena, as naturalists use these terms.

The Biological Society was organized for the purpose of considering and discussing the questions involved in the very broad-
est application of the term biology; in others words, organic
nature in any and all of her manifestations. Organized but
about two years prior to the death of Charles Darwin, it is not

surprising that its members have been very generally imbued with the spirit and interpretations which the illustrious author of "The Origin of Species" gave to the phenomena of life upon our planet. Not that they have been blind followers of the school which believes in the all-sufficiency of natural selection to account for life-phenomena; for a review of the communications and discussions and particularly of the addresses which have been delivered by my predecessors will show that in the search after truth, the ideas of Lamarck and others, who have pregnantly speculated on the philosophy of life, have been duly appreciated. Upon the one great question which, more than any other, has occupied biologists of late years, viz., whether functionally acquired characters are transmitted by heredity, there have been few more able contributions to the subject anywhere published than the papers and addresses of my distinguished predecessor, Prof. Lester F. Ward. Indeed, aside from the reasons already given, the choice of my subject to-night was in no small degree determined by an admission in one of his more recent and yet unpublished communications to the society, to the effect that the characters of neuters among the social insects offer the greatest stumbling block to the theory of the heredity of such acquired characters.

ORGANIZED INSECT SOCIETIES.

The social insects, or those which live in communities, and particularly those of the order Hymenoptera, which possess highly developed social characteristics, have, from the very earliest times, intensely interested the student of insect life. There are insects of other orders which are either social normally or become so by exception and for special purposes. Thus many Lepidopterous larvæ live together when young, but scatter when they grow older. In some cases there would seem to be no particular purpose in the association; in others, as in the common Tent Caterpillars (Clisiocampa spp.) the well-known Fall Web-worm of North America (*Hyphantria cunea*) and many similar species of other countries, the association is of a somewhat higher character, as the larvæ build a common web into which they retire at stated periods, and which helps to protect them both from the inclemencies of the weather and from the attacks of birds and other enemies. The highest development of this social trait in the Lepidoptera is found in the small Hyponomeutidae, and in a Mexican butterfly (*Eucheira socialis* Westw.)— the transformations taking place within the nest. The layers of silk in the last-named species are so tough that they have been used as parchment.

In one remarkable case among the Diptera, viz., in Sciara, a genus of small gnats, the larvæ have the habit of banding together in large masses, more or less elongate, all the individuals attached to each other, heads to tails, and the whole mass moving with one impulse and as a unit. They thus move across a road or field, like some huge snake, and are for that reason called "snake-worms," and really give us a very good illustration of how individual units may combine to make a compound whole. Many other insects have the exceptional habit of congregating together in large masses, but in almost every case the congregating is connected with undue multiplication and the desire to migrate to new regions. The habit is well exemplified in our notorious Army Worm, the larvæ of *Leucania unipuncta*, an insect which, over vast stretches of country, occasions great loss to our grain and grass crops by traveling from field to field and leaving devastation in its wake. Instances of this kind might be multiplied; but we do not apply the term social to such temporary associations of individuals, even where they have any specific purpose and are of annual recurrence. Nor do we apply the term social to those insects, of which there are many in different orders, which assemble together during the love or pairing season. The term is strictly confined to those species which permanently live together in colonies, and in which the social habit, with its consequent subdivision of labor, and differentiation of individuals, has become essential to their perpetuity.

BEES.

Living in such well organized communities, exhibiting so much intelligence, and yielding one of the most delicious sweets known, the Honey or Hive Bee has attracted attention from the earliest times, and ever since Aristotle, Virgil and Columella told what was then known of this industrious insect, it has been the subject of investigation. Honey and wax were far more important to man in olden time than they are to us who have so many substitutes for them, and the ancients gave much attention of the practical kind to bees. How very little they knew, however, of their true economy is shown by the prevalence of the belief that bees came from the carcasses of animals. This superstition as to the *Bugonia*, as exemplified in the biblical story of Samson (Judges XIV, 8) continued for twenty centuries and grew out of the resemblance to the Hive bee of *Eristalis tenax*, a Dipterous fly which breeds in putrescent matter. This fact, first clearly recognized by that excellent observer, Réaumur, has been fully established in a recent most interesting paper by Osten Sacken "On the so-called *Bugonia* of the ancients, and its relations to *Eristalis tenax.*" (Bullettino della Società Entomologica Italiana, Anno XXV, 1893). In fact the fabulous about bees prevailed till the beginning of the last century, when Maraldi, by the invention of glass hives, gave an impetus to correct observation, and led to the remarkable memoirs of Swammerdam, Réaumur, Schirach and Francis Huber.

The fact that the Hive Bee can be cultivated and controlled with a view to profitable industry, has served to heighten the interest in it, and since the invention in this country, in 1852, of the movable frame hive, by a retired ·clergyman, the Rev. L. L. Langstroth, progress in apiculture has been rapid and continuous. Of the more important subsequent inventions, many of them made in Europe but perfected in America, may be mentioned the honey-extractor, which, by centrifugal force, throws the honey from the comb, leaving the latter intact and ready to be used again; and the comb foundation, by which sheets of wax are impressed with the bases of the cells and employed to ensure straight and regular combs, to limit drone production and increase the honey product. With the bee-smoker in its modern form, bees are also much more easily controlled and manipulated than formerly. Much has been done, also, in ameliorating the races of bees, both by introducing races from other countries and by the crossing of these. There are some three hundred thousand of our citizens engaged in bee culture, and they add over twenty million dollars annually to the wealth of the country in honey and wax. This amount may be, and in the near future doubtless will be, very largely increased. It is, in fact, difficult to realize what an immense amount of honey is wasted from lack of bees to garner it, and the poet Gray would seem to

have had his own ideas on this subject when he wrote the familiar lines.

> "Full many a flower is born to blush unseen,
> And waste its sweetness on the desert air."

The service directly rendered to man by bees, however, in supplying the products mentioned, is but slight as compared with
the services indirectly rendered by cross-fertilization of our cultivated plants, and it has been estimated that the annual addition
to our wealth by bees in this direction alone, far exceeds that de-
rived from honey and wax. One of the latest discoveries bearing on this subject, very fully enforcing the general principle,
was presented to the Society for the first time within the past
year by our fellow-member, Mr. M. B. Waite, as a result of his
investigations for the Division of Vegetable Pathology in the
Department of Agriculture. He has proved that a majority of
the more valued varieties of our apples and pears are nearly or
wholly sterile when fertilized by pollen of the same variety, or
that they bear fruit of an inferior character and very different
from that produced when cross-fertilized; further, that were it
not for the cross-fertilizing agency of bees, scarcely any of these
fruits could be produced in the abundance and perfection in
which we now get them, and that to secure the best results and
facilitate the work of the bees, it is yet necessary, in the large
majority of cases, to mix varieties in the same orchard. Bees
were doubtless the earliest embalmers, since they use the propolis to encase and thus prevent the putrefaction of any intruder
which is too large for them to drag out of the hive.

There is much, even to-day, in the economy of the Hive Bee that
is yet debated among the best informed apiarians, but I will endeavor to give you an epitome of what is absolutely known of its
more important habits, structures and functions—the true life-
history, so to speak, of the bee. By going somewhat into detail
with this species, we may avoid repetition in treating of the other
social Hymenoptera, all of which have somewhat similar larvæ
and transformations. Let us, in imagination, proceed to an
ordinary well-kept apiary. Taking a bee-smoker in one
hand—one of the pattern invented by the late M. Quinby of
New York—we lift one corner of the hive cover or quilt, and
send enough smoke down among the bees to give them to
understand that they must submit to our manipulation. Draw-

ing out one of the brood combs, which is rendered easy by the movable frames, thousands of the bees are seen adhering to the surface of the comb. They are mostly workers, but in summer there may be seen numbers of stouter-bodied bees, which are the drones or males. If the bees have not been too much disturbed by the smoke or the removal of the comb, the queen may be seen walking slowly over the surface, surrounded by the workers, who, in deference, recede as she walks along, turning their heads toward her and advancing so as to touch her body with their antennæ. It was long thought that the queen exercises sovereign powers, and Shakespeare voices the popular opinion when, in Henry V, he says:

"They have a king and officers of sorts."

One of the earliest definitions of a queen bee in Webster's dictionary was, "The sovereign of a swarm of bees." In reality, however, the government of the hive is purely democratic. Each works for the common welfare, and only so long as the individual, whether queen, drone, or worker, is useful to the community, is it spared. With the exception of the drones, the queen is the only bee in the hive having the reproductive organs fully developed, and she is, therefore, the mother of the colony. During the more prolific season she lays two or three eggs in the course of a minute, and often as many as four thousand in twenty-four hours. Three days after deposition of the egg the young larva is hatched. It is the office of the younger workers, known as nurse-bees, to furnish these young larvæ with food, which they are assiduous in doing. In the case of the worker larvæ, five days suffice for full growth, when they nearly fill the cells. As with most other soft-bodied larvæ that are embedded in a semi-liquid nutritious medium, we find provision to prevent contamination of the environmental food with excrementitious matter. The food supply is, in the first place, highly nutritious, and nearly all capable of assimilation. Lest, however, any portion of the waste should enter the food, the larva is, according to Cheshire, rendered incapable of voiding anything during the time of feeding. The arrested development of the digestive system leaves the posterior inflection, which corresponds with the after bowel, unconnected with the middle bowel, and the slight accumulation of waste matter in this latter

is cast into the base of the cell at the last molt, and is covered
in the bottom of the cell by the lower part of the last cast skin
or pellicle, which also serves to line the rest of the cell and leave
it clean for the formation of the pupa. Thus, when the young
bee emerges, the cell needs but to be brushed out by the workers
to be ready to receive another egg or stores of honey and pollen
which are to form the winter food.

Just before pupation, or when the larva has acquired full
growth, the adult workers cover the cell with a convex lid composed not of wax alone, as in the case of the cappings of honey
cells, but of pollen and wax combined. The larva just before
pupation strengthens this cap by lining it with silk, which is also
slightly attached to the last cast skin. The pupa state lasts some
twelve days, and on the twenty-first day from the time the egg
was laid, the perfect bee cuts a circular opening in the cell
cap and makes its way out. The first care of this young bee is to
seek food from an open honey cell, and in the course of two or
more days it has acquired sufficient strength and consistence to
enable it to begin its labors as a nurse bee, doing for the developing larvæ what was so recently done for it. After a week's time
it takes short flights, noting well the location of its hive so as to
be able to return to it.

Queens are only bred when a colony is about to swarm, or
when an aged or failing queen needs replacing, or where an acci-
dent has deprived the hive of her services. If she be removed
from the hive during the working season, the bees are thrown
into great excitement, shown by the change of the contented hum
into one of alarm, by the hurried movements from the combs to
the entrance, and by the discontented flight to and from the hive.
If all the brood combs are removed the bees become panic-
stricken, and give utterance to a peculiar mournful note or distressed wail, quite different from the normal cheerful hum. In
time, however, this excitement subsides, as they become satisfied
of their loss. If the queen be returned, or a comb containing
young larvæ be introduced into the hive, the whole attitude
changes. The moment the first bee touches with its antennæ the
queen, or a comb, or any point over which she had walked recently, it sets up a loud and cheerful hum, and the occupants of
the hive, even those unable to see the comb, immediately catch
the sound, and crowd toward the point whence it first pro-

ceeded, repeating the jubilant note. If only a comb of larvæ be given them, they still recognize it as a deliverance from the threatened extinction of the colony. In a few hours one of the cells over a larva two or three days old will be enlarged by the partial destruction of the walls of the adjoining cells. This enlarged cell is built outward and downward, and the larva is fed on the so-called royal jelly or bee-milk. The supply of this food is always plentiful, and when a well-developed queen has issued, it is not uncommon to find a quantity of the food, in a partially dried, jelly-like mass, in the bottom of the cell. When, preparatory to swarming, young queens are being reared, the workers have to guard them, even in the cell, from the jealous fury of the reigning queen, and the instinctive rivalry and conflict between queens, accompanied by a peculiar shrill battle-cry, first noticed by the elder Huber, are quite suggestive of similar conflicts between rival queens in human monarchies.

Economy of Hive. Social Organization. Division of Labor.

Each bee, as already stated, labors for the good of the commonwealth of which it is a member. Of them it might well be said:

"Salus rei publicæ suprema lex."

It is the welfare of the colony which directs the actions of all, and not the will of the queen. Indeed, it would seem that the latter performed her important function—that of supplying the hive with eggs—only when the workers willed it, their own condition of prosperity as regards stores, or their anticipations of the future needs of the colony as regards population, causing them to supply the queen liberally with food rich in nitrogen—a partially digested substance or a gland product, or perhaps a mixture of both, which she alone cannot produce, yet without which any considerable production of eggs is an impossibility.

As Evans remarks:

"The prescient female rears her tender brood
In strict proportion to the hoarded food."

We must, then, credit the industrious and provident workers with the chief influence in shaping the policy of the hive. They are the *servum pecus*—the living force—of the colony. And to the end that order and efficiency of effort may prevail, they have, we find, a marked division of labor. In the normal condition of

the hive the young workers, as already stated, care for the
brood—a labor which they take upon themselves within two or
three days after issuing from the cell. The glands which secrete
a part of the food required by the developing larvæ are active
during the earlier part of the life of a worker. Later these
nurses become incapable of doing their work well, as the gland
system becomes atrophied. When a few days old they take
short flights, if the weather favors, but seldom commence
gathering stores before they are fifteen days old. Wax production is more essentially a function of the workers in middle life, and it is particularly noticeable that those bees fashion-
ing the wax into combs are principally of this class. Many of
those acting as foragers do, however, secrete wax scales, which
are doubtless, in the main, utilized. Among the outside workers
and hive-defenders some bring honey only on certain trips or for
a time; others honey and pollen; others water,'and yet others
propolis or bee-glue to stop up crevices and glue things fast.
Meanwhile some are buzzing their wings at the entrance to ven-
tilate the hive, and others are removing dead bees, dust, or loose
fibres of wood from the inside of the hive or from near the entrance, or are guarding this last against intruders, or perhaps
driving out the drones when these are no longer needed.

SWARMING.—Perhaps there is no action on the part of the
Hive Bee which more distinctly indicates its intelligence and
power of communication than the act of swarming. The fact
that queen brood is being reared in the hive is the best evidence
that the colony is preparing for flight or swarming; but, in addition, it is noticeable that on the day of swarming the whole
colony is excited, and in a measure has abandoned ordinary
duties. For days previous to the event, scouts have been searching for a favorable hollow or crevice or place in which to house
the new colony, and when the time finally comes, which is
usually in the hotter part of the day, all the individuals of the
hive leave after the peculiar preparatory flight around the hive,
known as swarming. The impulse to leave is such that many
individuals not yet capable of flight, fall to the ground, and the
hive is practically abandoned by all those within it at the time
of swarming. Individuals alight on some bough or object near by,
with a view primarily to organization and the sending out and
return of additional scouts. During this period a cluster will

remain more or less in repose, but when once the location for a permanent dwelling has been finally determined upon, the whole mass will leave as with one impulse and fly swiftly and directly to the new home. With the first swarm that the new colony sends out it is the old or fertile queen that goes with the new swarm, but with the after swarms, which issue in about a week, it is a virgin queen that accompanies. The old colony begins again with the few individuals unable to follow the departing swarm, and which have crept back to the old hive, with those which at the time of swarming were busy in the field, and with those which issue from the yet undeveloped brood.

It is a popular mistake to suppose that mating takes place during swarming. If a virgin queen goes with the swarm, she subsequently takes the nuptial flight from her new home. As she flies swiftly and strongly, only the strongest and most vigorous drones are able to mate with her, and there is every opportunity for cross-fertilization with drones from some other colony. It has also been noticed that drones have a way of congregating in some particular spot, as though awaiting their chance of thus mating with the queen.

The more important special Organs.

The different structures and organs of the Hive Bee are most interesting, but I can allude only to a few of the more striking. The tongue is a very complex organ, fitted for obtaining minute quantities of nectar from the flowers that secrete it but sparingly, or to remove the same substance rapidly when found in abundance. The figure of the head and appendages thrown on the screen will illustrate this organ in detail. We have here the mandible, mostly used for cutting and moulding the wax, the maxillæ with their palpi, the labium and labial palpi, and finally the ligula or true tongue with its spoon-like tip. This is extremely flexible, and consists of a rod or central portion, nearly surrounded by a sheath which is covered thickly with hairs, which aid, by capillary attraction, in taking up the liquid food. A lapping motion, when the liquid is abundant, causes the liquid to be lodged among the hairs of the tongue, which can be partially drawn into the mentum, and from this point the maxillæ above and the labial palpi below unite to form a tube around it, which is closed above the extension of the epipharynx, and by alternately arching and depressing the maxillæ, the space enclosed is increased or decreased, thus producing suction and drawing the liquid held on the tongue into the opening of the esophagus.

When drawn from the flowers the nectar is thin and watery and lacks the qualities of the delicious honey into which we find it converted when removed from the cells sealed by the bees. This watery substance is evaporated to the proper consistency in the heat of the hive and by currents of air passing over the surface of the combs before the cells are sealed, these currents being created by bees stationed at the entrance and buzzing incessantly. There has been much discussion among apiarians, as among writers, as to whether the bee gathers or makes honey. Strictly speaking it does both. Formic acid is contained in the blood of the bee and especially in the salivary glands, as recently demonstrated by von Planta of Zurich, and when the gathered nectar, which easily ferments, is regurgitated from the first stomach into the cell, it is combined with sufficient formic acid to change the cane sugar into invert sugar (dextrose and levulose in equal proportions) while the evaporating process just described eliminates the superflous water; so that honey which resists fermentation is essentially a made product.

I would also draw your attention to the wax-producing organs (See Fig. 3a, *a*). If we examine the underside of the abdomen of the worker, the exposed portion of each segment will be seen to be covered with a web of hairs, and by elongating the abdomen, each segment, with the exception of the first and sixth, is seen to bear two shallow, irregularly-shaped plates, one on each side of the median ridge, which is extended as a rim around the whole contour. These pale yellow, smooth plates are in reality wax moulds, the wax glands being under the plates and the secreted wax reaching the surface by osmos through the thin membrane and hardening into a somewhat brittle scale, resembling in appearance a minute, nearly transparent fish scale. The wax is secreted under conditions of great heat, the bee ascending for this purpose to the top of the hive, and the wax producers consuming a large amount of honey.

The next structure of importance to which I would call your attention is the wax pincers (Fig. 1b, *a*, *b*), which is a modified structure of the juncture of the tibia and metatarsus of the pos-

terior legs. With these pincers the wax producer plucks a scale
from one of its wax plates, passes it rapidly forward to the
mouth, and here makes it plastic and at the same time more or
less yellow, by continually manipulating and chewing it between
the mandibles. Then the bee sticks it to the under surface of
the hive cover or object to which the comb is to be attached.
More wax is added, forming a slight ridge, which is chiseled
or pressed from each side by workers, using their firm and highly
polished maxillæ, and placing themselves so that their range of
work will overlap just one-half. As this ridge is built down,
forming a sheet—the septum upon which the cells are constructed—the sides of the latter are started simultaneously. In
their efforts to make the cells concave at the bottom and so as to
fit together at the sides without loss of material, mutual pressure
results in straight lines, the sides becoming hexagonal in outline,
just as six soap-bubbles resting against a seventh cause the latter
to assume a hexagonal form; while the bee starting a cell on the
bottom of one already commenced on the other side, naturally
takes the apex of the latter as a part of the boundary of its own
cell in order that the latter may also be concave. Thus three
rhomboidal faces forming the base of one cell, form individually
a part of each one of three cells on the opposite side.

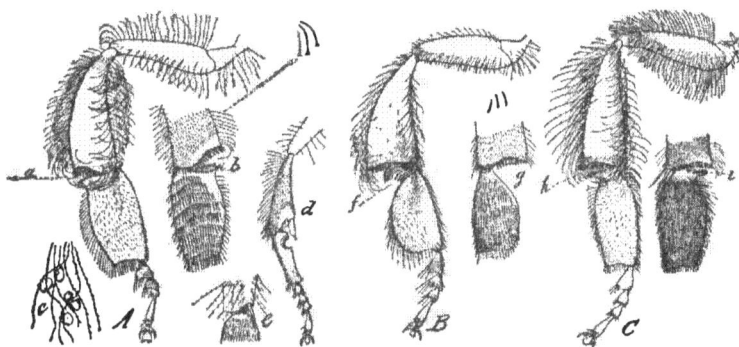

FIG. I.—MODIFICATIONS OF THE HIND LEGS OF DIFFERENT BEES : A, *Apis* : *a*, wax
cutter and outer view of leg ; *b*, inner aspect of wax cutter and leg ; *c*, compound hairs;
d, anterior leg, showing antennal scraper. B, *Melipona* : *f*, peculiar group of spines at
apex of tibia ; *g*, inner aspect of wax cutters and first joint of tarsus. C, *Bombus* : *h*,
wax cutter ; *i*, inner view of same and first joint of tarsus—all enlarged. (Original.)

Finally I would call your attention to the arrangement of
the hairs on the inside and outside of the legs (Fig. 1, A),

so well fitted for collecting and holding pollen, and to what is known as the antenna-comb or strigil (Fig. 1, *d*), a structure with which the bee cleanses itself, and especially the antennæ, which are organs of extreme sensibility and need to be kept well cleaned. This structure occurs on the underside of each front leg and is a semi-circular cavity in the upper end of the metatarsus. The cavity is fringed with stiff hairs or spines, forming a comb. The distal or opposing end of the tibia is furnished with a spur, slightly concave on the inner surface and known as the velum When the tibia and metatarsus are bent at right angles, the velum falls over the cavity and forms an almost circular opening just large enough to snugly hold one antenna.

These are the more conspicuous structures, though there are others of minor importance, all indicating remarkable adaptation to special purposes and to the necessities of the bee.

The Hive Bee is but one of many species of its family, and while representing the most highly organized of the social insects, has many cousins and more distant relatives which are equally interesting. The numerous bees, with their diversified habits, have an especial interest, when studied structurally and biologically, as throwing light on the origin and development not only of the higher social habits and intelligences of the true Hive Bee, but also of its structures, so remarkably fitted for their special purposes.

Species of Genus Apis and Variations in Apis mellifica.

The old conception of the Hive Bee, its attributes and structures, was that it exemplifies in a marvelous manner creative wisdom for man's interests.   Yet while it represents great perfection of organization and of structure, for particular ends, this perfection is relative and not absolute.   Though a number of species of the genus Apis have been characterized by authors, there are but four well defined species so far known, and three of them— *A. dorsata, A. indica* and *A. florea* are confined to India and the East Indian and Philippine Islands.   The fourth, *Apis mellifica,* or the common Hive Bee, was originally introduced into this country from Europe, and doubtless had its origin in some parts of Asia.   It has followed civilized man in his migrations over the globe, and has frequently anteceded him, and, being semi-domesticated, has been more or less influenced by him, as have other

domesticated animals. Some ten different types of the species
have been characterized by specific names, two of them—viz.,
adansoni Latr. and *unicolor* Latr.—being considered good species
by Fredk. Smith, while a still greater number are recognized by
local names among apiculturists. These varieties and races show
every variation in color through the various shades of black, gray
and golden-yellow, as also every variation in disposition, industry, and tendency to swarm, and especially in honey-gather-
ing proclivities. (See Note 1.)

Of the East Indian species only one, *Apis indica*, is cultivated.
This bee, which is considerably smaller than our own, building
smaller combs composed of smaller cells—36 to the square inch—
chooses when wild, a hollow tree or rocky cavity for its home.
It is kept to a limited extent by the natives, earthen jars being
used for hives, but the yield of honey is small.

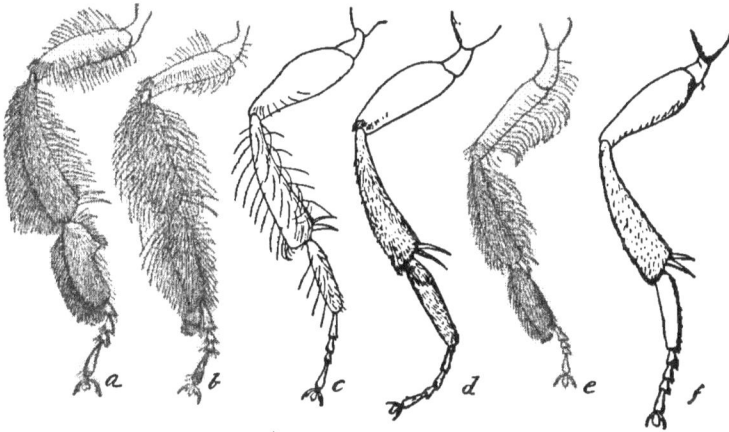

Fig. 2.—Modifications of the hind legs of different bees : *a*, Anthophora ;
b, Melissodes ; *c*, Perdita ; *d*, Nomada ; *e*, Agapostemon ; *f*, Nomia—all enlarged.
(Original.)

Apis florea, the smallest of the genus, with slender, orange-
banded body, builds in the more open country of India, attach-
ing a single tiny comb to the twig of some small shrub. The
worker cells are 81 to the square inch of surface, the drone
cells 36.

Apis dorsata, the Giant Bee of India, attaches its mammoth
combs to the limbs of tall forest trees or to overhanging ledges

of rock, generally building a single comb as much as six feet long and two or three feet wide. Great quantities of wax and honey are obtained from this bee by the bee-hunters in India and the islands southeast of Asia. It has not been permanently domesticated; nor is it certain that it can be. The workers of this species are about the size of the queens of *Apis mellifica*, or from seven-eights of an inch to an inch long. The bodies of the bees are slender and wasp-like, and beautifully marked across the abdomen with bright orange bands. (See Note 2.)

While the different species of the genus Apis thus differ in size, coloration, temperament and habit, there is comparatively slight variations in structure; a necessary inference for every zöologist. But if we study the other species of the family Apidæ, we shall find every variation, and obtain a very good idea of how the special organs in Apis may have been evolved and perfected from simpler organs in other genera. This may be illustrated by a few sketches of some of the more important structures, as for instance, the polliniferous organs and the wax producing apparatus. (See Figs. 1, 2 and 3.) The figures already thrown on

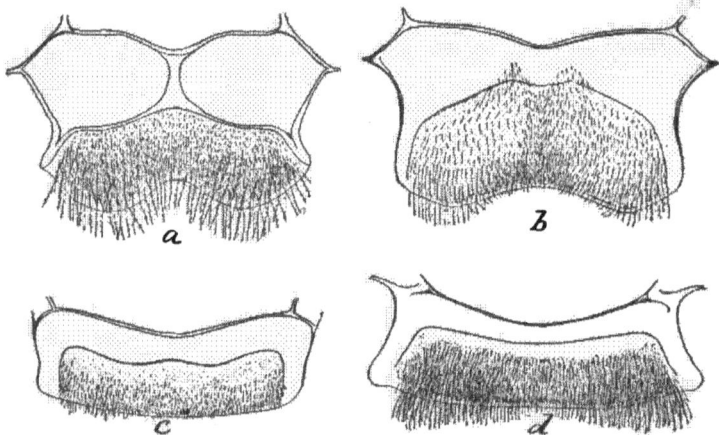

FIG. 3.—WAX DISCS OF SOCIAL BEES: *a*, Apis worker; *b*, Apis queen; *c*, Melipona worker; *d*, Bombus worker—all enlarged. (Original.)

the screen very well illustrate the fact that the modification of structure and hairy vestiture, which facilitate the collection and transportation of pollen, while exhibited, perhaps, in the greatest perfection in the Hive Bee, is nevertheless an evolution from similar structures possessed by other species of social bees, such as the Meliponæ and Bombi, and still more remotely from such as are possessed by the solitary bees. Here again I trust to diagrams, and relegate detailed exposition to a note. (See Note 3 and Figs. 1 and 2.)

In the production of wax the Hive Bee exhibits a lavishness not found in any of the wild bees, not excepting the species of Trigona and Melipona, which approach it most nearly in social economy. As a result we find that the wax-secreting organs of Apis are much larger than in any other wax-producing bees. In Bombus they are greatly reduced and otherwise different in structure, resembling, however, very closely, those obtaining in Melipona and Trigona. In the solitary bees, which produce no wax, these specialized structures are entirely wanting. (See Note 4.) But the most interesting fact is that in the queen bee, in which they are functionless, they are nevertheless present, but more nearly resemble the same structures in Melipona.

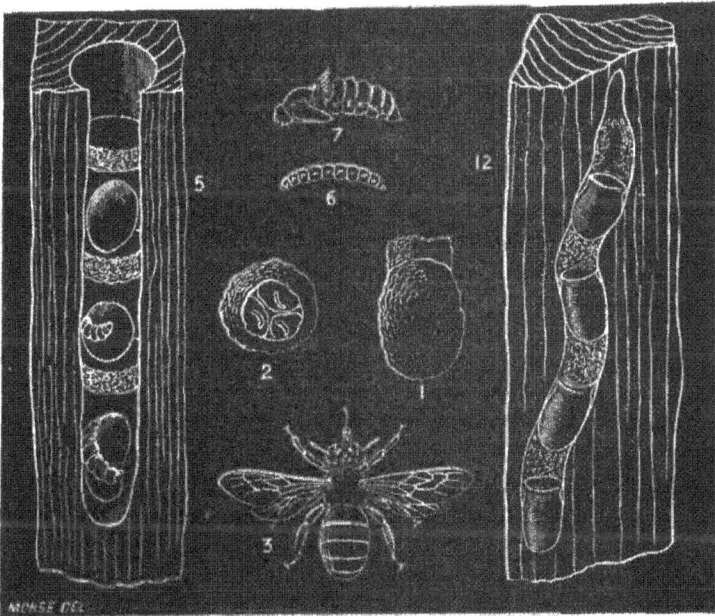

FIG. 4.—ARCHITECTURE OF BEES: 1, cell of bumble-bee; 2, end of same showing eggs; 3, Xylocopa virginica, the carpenter bee; 5, cells of same; 6, larva of bee parsi e, *Anthrax sinuosa*; 7, pupa of Anthrax; 12, cells of mason bee, *Osmia lignivora*—natural size. (After Packard.)

The architecture of certain solitary bees is shown in Figs. 4
and 5.   These solitary bees, no matter in what situations or of
what material they make their cells, generally store them with
honey or pollen, and after depositing an egg, cap the cell and
leave the young larva to care for itself.   The habits of the social
Bumble-bee (Bombus) are but a step in advance, as the larvæ are

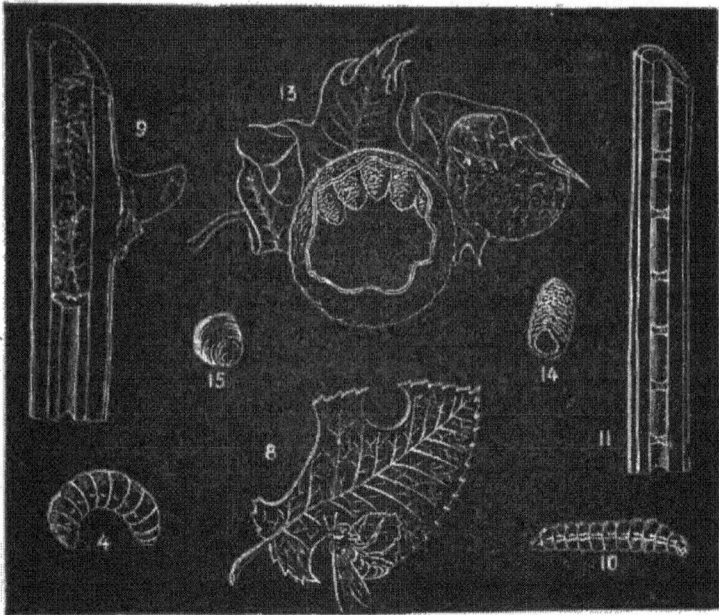

FIG 5.—ARCHITECTURE OF BEES (continued) :   4, larva of *Xylocopa* ; 8, leaf-cutter bee,
Megachile ; 9, cells of *Megachile* in elder ; 10, larva of upholsterer bee, *Ceratina dupla* —
enlarged ; 11, cells of same in elder, ; 13, cells of *Osmia s.militma* in deserted oak-gall ;
14, earthern cell of same ; 15, pollen mass of Osmia—natural size   (After Packard.)

developed in a mass of pollen and honey, in which they form
rather imperfect cells.   When full grown each spins a silk cocoon
which is thickened by a certain amount of wax, which is added by
the adult bees.   The females labor and several co-operate in the
same nest.   In the Bottle-bees (Melipona) a still further step is
seen, as the cells, of a rather dark, unctuous wax, are formed into
. regular combs and are somewhat imperfectly hexagonal.   They
are, however, in single horizontal tiers, separated and supported
by intervening pillars, more like the nests of the social wasps, and
the cell is sealed after the egg is laid upon the stored food, just as
in the case of solitary bees.   The honey is stored in separate flask-
like cells, and but one queen is allowed to provide eggs.

SOCIAL WASPS.

The popular conception of these interesting insects is decidedly at variance with their deserts. Wasps are generally considered as thieves, robbers, idlers and vagabonds; as impertinent and inquisitive, invading our homes and devouring anything and everything their fancy craves, as sugar, fruit, meat, wines, etc., and resenting any interference in such a pointed way as to bring pain and rage to the incautious or meddlesome individual who interferes with their operations. The term "waspish," one of the most expressive in the language, very well denotes the popular feeling towards these somewhat maligned insects. Granted that toward other insects they are cruel, and that they courageously resent interference, yet the fact remains that they are seldom, if ever, the original aggressors in the infliction of punishment, except in the capture and appropriation of other insects as food— a course which finds its counterpart in every other carnivorous insect or higher animal, and is justified even by the example of man himself. In their relationship with each other, the wasps are polished and gentle, and never quarrelsome so far as their own species are concerned; and they never turn robbers or marauders of their own kind, as do the more lauded bees, among which we have what are known as the corsair bees, which frequently rob their sisters of the sweets and pollen which they have collected with great pains and indefatigable industry. These robbers even lie in wait, and scheme and plan in bodies for the success of their raids, as do thieves among men. Wasps never resort to such cowardly proceedings, and hence strictly speaking, are not robbers at all; for aside from their own kind the world is their legitimate pray.

The family Vespidæ, to which the wasps and hornets belong, comprises some thousand known species. They closely resemble bees, but differ in possessing more cylindrical bodies with a harder, smoother integument. The wings are longer and folded once longitudinally, and when at rest are laid flat on the body. The antennæ are elbowed, and the jaws are large and powerful. Their eggs are at first nearly spherical, but rapidly become ovoid. Their larvæ, as in the other social Hymenoptera, are legless and helpless grubs, entirely dependent on the adults for food and care. The family comprises two natural groups, viz., the Social

Wasps, having, as with bees and ants, three forms—males, females, and workers or neuters; and the solitary species, in which only females and males occur.

The common Bald-faced Hornet (*Vespa maculata*) is a familiar example of the first-named group. It constructs remarkable nests of various patterns, of a gray, paper-like material, and suspended to the branches of trees and shrubs, or to the rafters of houses. In the second group, on the contrary, the species construct cells or nests, consisting usually of single cells, of sand or mud, in protected situations; store them with insect food for the larvæ, and then abandon them altogether. The former— "natural paper-makers from the beginning of time," as Harris properly styles them—have always done what man, with all his boasted superiority, has only in recent times learned to do; viz., make paper of wood. They resort for this purpose to such woody surfaces as have long been exposed to and bleached by the action of the elements. With their powerful mandibles they tear off minute filaments and chew them into a fine pulp, which they afterward spread into a thin sheet of strong, water-proof paper, out of which they construct their nests. These nests are of two kinds, one made by the true Vespas, as in the case of the Bald-faced Hornet just alluded to. Here the outer covering forms a more or less regular globose body, with a single circular orifice at the bottom, the combs being arranged within this covering in horizontal tiers or stories. In the second category we have the nests of the wasps belonging to the genus Polistes, which are more particularly known by the name of paper wasps. Here the nest has no outer envelope, and is usually limited to a single tier of cells suspended by one or more peduncles or short stems. They are usually attached in the open air to the branchs of trees, or are fastened to the underside of the rafters of porches, etc., garrets being favorite places for their construction. Some of the hornets, such as the "yellow-jackets," are found occupying the deserted nests of mice, suspending the tiers of cells from the ceiling and lining the burrow with a layer of woody paper. The burrows are enlarged from time to time as the growth of the colony requires additional space, and in late autumn are often found large enough to fill a bushel measure, containing sometimes from 15,000 to 20,000 cells. In all these cases the tiers of cells are attached to each other or to other supports by strong pillars of the same *papier mâché* material, but of darker color and firmer texture.

The combs of these paper wasps and hornets are not double, as in the case of the Hive Bee, and the cells, which are less perfectly hexagonal, have the mouth beneath and are in horizontal instead of vertical layers. They differ from the cells of bees, also, in that they are used solely in the reception of the larvæ and, except in some tropical species,* not for the storage of honey or pollen. The nests of wasps vary greatly in the different species, and find their greatest perfection in the card-making species of Cayenne (*Chartergus nidulans*) the outer covering of which is nearly white and as tough as the stoutest card-board.

The life-history is very interesting. Perfect females or queens and males are produced in the autumn, in cells of large size, and in the case of the hornets proper, these are developed in the lowest and last constructed of the cells. The males and the workers or imperfect females, perish at approach of winter, while some of the fertile females hibernate in sheltered situations. These, in the following spring, originate new colonies, and may be seen about early spring flowers, which they frequent for honey, but more particularly to prey upon other insects attracted to the blossoms. Singly and unaided they originate the new colony, building cell after cell, supplying each with an egg, and persistently bringing home food for the growing young. All these cells in the early season produce neuters or working females only. These, as soon as developed, assist the hibernated mother or queen in the enlargment of the nest and the care of the young. She, after having once started her colony, rarely leaves it, but remains and devotes herself solely to the duty of egg-laying. The workers become by far the most numerous, and by late summer are everywhere found moving actively about in search of food for the home brood. They are less than half the size of the perfect females, and considerably smaller than the males, which are easily distinguished by their more slender bodies and very long antennæ. The males are not mere idlers, as in the case of the bees, but occupy themselves with various labors about the nest, and while the male bee is in the end ruthlessly destroyed

*St. Fargeau states that he has often, in *Polistes gallica*, found cells filled with honey.

by the indignant workers, the male wasp is respected and protected, and dies a natural death.   In the large nests of hornets,
the number of males and perfect females produced in the autumn
amounts to several hundred, and of these comparatively few
females successfully hibernate.   Were it otherwise ordained, these
insects would become too numerous for the comfort of the rest
of the world.

The larvæ are fed from day to day with a prepared liquid food
which is disgorged from stomachs of the adults.   These
prey upon other insects, and also feed upon animal or vegetable
matter to which they have access, and are particularly fond of the
sweets of fruits, melons, etc., also of sugars and honey, all of
which are eaten greedily, and commingled and prepared in the
stomach as food for the young.   Wasps are not particularly active
themselves in the collection of honey from flowers, but are very
prone to rob the hives of bees whenever opportunity offers.

We have seen that in the case of the Hive Bee the unfertilized
egg, including the egg deposited by the worker bee, invariably
produces a drone or male.   The experience of English observers,
indicates that the reverse of this is true of the social wasps,
and that, instead of males being produced from eggs of workers
or non-fertilized wasps, other workers similar to the parent are
produced.   Thus from nests from which the queen wasp is
removed quite early in the spring, the generation of workers
continues through the season as freely as if the queen were still
present to lay eggs, showing that the brood is kept up by the
progeny of workers having no access to males, which only appear
in the fall.   Leuckart has also shown, by careful dissections, that
nearly fifty per cent. of the worker generations in the latter part
of the summer at least, have fully developed and developing eggs
in their ovaries. .

It must be noted, however, that the experience of Von Siebold
with *Polistes gallica* directly contradicts the observations of English investigators.   His experiments carried on in precisely the
same way, indicate that, with this species at least, the eggs from
the workers produce males.   There would, therefore, seem to be
no uniformity in this regard among the different species of the
family, both arrenotoky and thelytoky occuring among them, and
possibly in the same species at different seasons.

In the case of Vespa there is no difficulty in separating the

fertilized autumnal queen from the worker generations, the former being considerably larger and presenting even more marked
differences from the worker than occur in the similar states of the
bee. With Polisites, however, the difference between the fertilized
queen and the summer broods of workers is much less marked,
and it is more difficult to distinguish them. The abdomen
of the true queen of Polistes is somewhat longer and larger than
that of the worker, but the variation is so slight that accurate
separation is usually impossible, and there is probably less
difference between the worker and the fertilized female than ob-
tains with the social bees, the worker being quite capable, in
many cases at least, of producing eggs which will develop into
other workers, and at the proper season also, doubtless, into males.
The distinction between the summer broods and the autumnal
females which are fertilized and hibernate, is probably produced
by food conditions, as in the case of bees, although accurate observations are wanting.

Just as in the case of bees, the study of the wasp family (Vespidæ)
in its different genera and species, reveals every gradation in habit,
from the solitary species to the more highly organized or social
forms, and these differences in habit are accompanied by differences in structure, so that the origin of the higher or more social
forms may be traced through the less specialized.

Many instances might be cited in illustration of the great intelligence of wasps, and especially in proof of their wonderful
sense of direction. On the whole they exhibit a rather higher degree of intelligence than do the bees, in the remarkably varied pro-
visions which they make for their young. Their habitations, also,
complete in themselves, and built chiefly of extraneous matter
not secreted from their own bodies, indicate greater architectural
dexterity than is found in the bees.

ANTS.

Few insects have attracted more attention, or have become more
renowned than the ants. Considering their comparatively di-
minutive size, their endless activity, and the wonderful results
they accomplish, this is not to wondered at.

Up to the present time some fifteen hundred species of ants
have been described, the great majority of the species, as well as
the largest and most rapacious, occuring in tropical and semi-

tropical countries. Some two hundred species have already been described from North America, many of which are nearly related to or even identical with those of Europe; while some are cosmopolitan, having been distributed by the agency of man over almost every part of the world. One of the best known of these cosmopolitan forms is the the little Red Ant, *Monomorium pharaonis* Linn., a grievous household pest. Under the tribal term *Heterogyna* Latreille, the ants are divided by the later systematists into four families (by some considered sub-families), namely, the Formicidæ, the Poneridæ, the Dorilidæ and the Myrmicidæ. The first family, Formicidæ, comprises all those species which are destitute of a sting, except in the genus Œcophylla, and are further characterized by having but one node or scale connecting the abdomen with the thorax, and by the habit in the larva of constructing for pupation, a dense, smooth, ovoid, silken cocoon. The remaining families are possessed of a sting, the Poneridæ agreeing with the Formicidæ in the cocoon-forming habit of the larva and in having but a single node or scale connecting the thorax and abdomen, but having an additional, more or less pronounced constriction between the first and second abdominal joints. The Dorilidæ are somewhat aberrant, the female and worker, so far as known, being blind, and nothing being yet known of their larvæ. In the last family, the Myrmicidæ, there are two well-developed, freely mobile nodes between the abdomen and the thorax, and the larvæ are unprotected by any cocoon during pupation. The most interesting and destructive species occur in this family.

Let us glance briefly at some of the species, more according to habit, however, than this classification, and preferably our North American species. Thus they may be considered as Carpenter, Mound-building, Harvesting, Honey, Leaf-cutting, Nest-building and Driving or Foraging ants. (Note 5.)

Ant Economy and Habits.

ANT WARS.—Very many most interesting accounts of the intelligence and battles, and of the curious persistency of ants, especially of the foraging species, are recorded by travellers in tropical countries, and particularly by the late Henry Walter Bates in his "Naturalist on the River Amazons". It is a well established fact that ants, like human beings, do at times declare war against

other species, or even against colonies of their own, while with
many species there is a form of neuter known as the soldier which
seems to be developed for no other purpose than to defend the
colony or make war upon some other colony. The soldiers are
characterized by an enormous and abnormal enlargement of the
head, jaws and mouth-parts. In these wars the greatest pugna-
city and courage are exhibited, the contest lasting sometimes for
days, and the weaker party ultimately succumbing from sheer
exhaustion and decimation.

There is a gradation in the warlike spirit in different species
and genera. Thus in Myrmecina and Tetramorium the ants do
not fight, but roll up and feign death. Lubbock shows that in
Formica exsecta, an active but delicate species, the individuals
advance in serried masses, and that when fighting with larger
species, like *Formica pratensis*, several in unison, attack an individual of the latter, some of them jumping onto the back of the
foe and sawing off the head from behind. The species of Lasius,
he says, will suffer themselves to be cut to pieces rather than let
go when they have once seized an enemy, while *Polyergus rufescens*, the notorious slave-making ant of the Amazons, seizes the
head of her enemy by closing the jaws, so as to pierce the brain,
thus paralyzing the nervous system; so that a comparatively
small force of Polyergus will fearlessly attack much larger armies
of the small species and suffer scarcely any loss themselves.

SLAVE-MAKING.—Nor must I pass without brief mention of another fact which has been well observed among ants, namely,
that some of the species repeatedly raid the colonies of weaker
ants and make slaves of them. In most cases it is a large pale
ant which enslaves a small black ant, and this is done either by
capturing fully developed workers or more often by carrying
home from the weaker colony larvæ and pupæ and allowing these
to develop in the formicaries of their masters.

It is most interesting to note, also, that the slave-making habit
among ants produces the same demoralizing results for the slave-
maker that it does among men. The habit is degrading. Thus,
as Lubbock points out, *Polyergus rufescens* has become entirely
dependent on its slaves. It has lost the power of building, as
also most of its domestic habits. Its impotence away from its
slaves has gone so far that even the habit of feeding has been lost,
and it will starve in the midst of plenty rather than feed itself.

Such cases as this, of an animal having lost the instinct of feeding, are extremely rare in nature, but the habit here has even
affected the structure, for the mandibles of the slave-makers have
lost their teeth and are useless except as weapons of war.

BURIAL GROUNDS.—It would seem almost incredible, but there
is nevertheless good evidence that some species of ants habitually
form burial grounds for the dead. An esteemed friend and re-
liable observer, Mr. Henry G. Hubbard, informs me that he has
carefully studied the habits of a black mound-making ant in
Montana, (*Formica subpolita Mayr*), the mounds being made in
dry situations in the mountains. There are always burial pits
just outside the hill, connected with it by passages; and these
burial pits contain generally a double handful of dead ants, with
occasional fragments of other insects. They are made in firm,
hard soil, and consist of a clean neat chamber, sometimes as large
as one's two fists. In moist ground the same species of ant does
not seem to use the same method of burial. These facts are all
the more interesting as showing how the same species may develop
a local habit, as *subopolita* is now considered but a variety or sub-
species of the widespread *F. fusca* L.

FOOD-HABITS.—In Note 5, in speaking of the several species,
I have recorded in detail some food-habits of our ants. Taken
as a whole they are truly omnivorous, feeding upon all sorts of
plant and animal matter, storing various kinds of vegetation, and
even, as in the case of the leaf-cutter ants, cultivating certain
fungus growths for food, but particularly relishing the sweets
obtainable from plants and other sources, and more especially
from the excrementitious and other secretions of plant-lice and
bark-lice.

KEEPING AND RAISING KINE.—There is no work upon ants
which does not refer to their well-known habit of guarding and
encouraging plant-lice, protecting them from their enemies, and
in other ways looking after their welfare. This attitude toward
various species of Aphididæ is essentially selfish, as these,
when carressed, yield a sweetened liquid which the ants much
covet. For this reason the Aphides have been denominated, in
popular parlance, the ants' milch-cows. Certain species of plant-
lice are frequently attended by particular species of ants, and
there is often a remarkable colorational harmony between a particular ant and the Aphidid colony which it cherishes. It is not

generally known, however, that the ants do more, and show an exceptional intelligence in carrying the eggs of the plant-lice in autumn into their own formicaries, bringing them together in little heaps and taking every precaution to preserve them through the winter. These eggs are carried back in spring to the plant upon which the particular Aphidid is nourished. There are, moreover, a number of other insects which the ants foster in their homes and from which they obtain coveted secretions; so that they may be said to utilize various kinds of cattle.

EARLY STAGES OF ANTS.—The transformations of ants are similar to those of other social Hymenoptera where the young are fed and cared for by the workers or nurses. The eggs are, as a rule, deposited by what may be called queens, i. e., by females more highly fed and developed than the rest, and devoted solely to the propagation of the species. It has also been noted that, in an emergency, where the females have perished, eggs may be deposited by the workers, as in the case of the Hive Bee, and also, as in that case, that these unfertilized eggs produce males only.

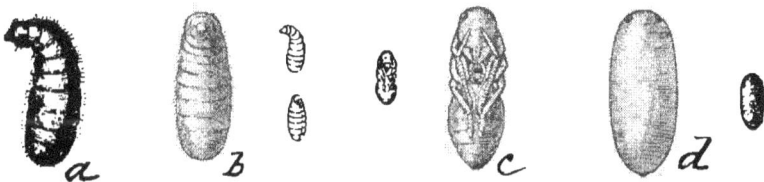

FIG. 6.—DEVELOPMENT OF FORMICA RUFA : *a*, larva, lateral ; *b*, do., ventral view ; *c*, pupa ; *d*, cocoon—enlarged, the outlines showing natural size. (After Dalton.)

The eggs are yellowish-white, ovoid or oblong-ovoid, very delicate in texture, and require from two to three weeks, or longer, for hatching, according to seasonal conditions. The larvæ are soft, white, legless grubs, having no eyes and being perfectly helpless. The small head is curved down on the breast and provided with but rudimentary mandibles. There is at first no apparent difference between the larvæ destined to produce the different kinds of individuals, but the growth of those destined to become workers suddenly ceases, whereas that of those destined to become perfect females, continues. As in the case of the larvæ of the bee, the workers are therefore but arrested or undeveloped females, and there is every reason to believe that the ultimate organization is a result of a difference in the kind of food or amount of food supplied by the nurses; so that practically the constitution of the formicary is regulated by the colony itself. The helpless larvæ and pupæ are moved from place to place, and most tenderly cared for by the nurses, which understand the requisite conditions of warmth, fresh air, protection against cold, rain, and other injurious influences, and which feed their young charges with a liquid discharged from the mouth, very much as in the case of the bee.

While the mandibles are used for tearing all sorts of substances, it is the juices of these which are lapped up by the tongue, and which can be regurgitated from a fore-stomach or pouch, in order to feed the young and the queens. These young are, also, arranged by the workers in groups of different sizes and ages, with a view to regulate the amount of food necessary for each stage. The larval life varies very much, so far as observations have been made, as its duration may extend from six or seven weeks to several months, according to the species. Some species even hibernate in the larva state. I have already indi-

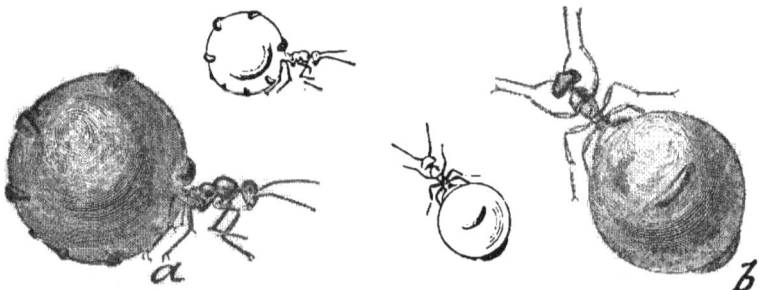

FIG. 7.—HONEY ANTS: *Myrmecocistus mexicanus*; *a*, side view; *b*, from above—enlarged, the outlines showing natural size. (after Lubbock.)

cated the differences in habit as to the formation of a cocoon or pupation without a cocoon, in the different families of the group; but a difference is noticeable in this respect, even in the same formicary, as first observed by Latreille. Those which pupate in cocoons are often unable to extricate themselves when mature, and are then tenderly assisted by the workers, who also aid in the unfolding of the wings, and cleansing of the newly-developed ant. (Fig. 6, shows a typical larva, nymph and cocoon).

The individuals of the formicary are therefore composed (1)

of neuters or workers, which are all females arrested in development; (2) of males; and (3) of fully developed females or
queens. All the males and females acquire wings, which are, however, torn off after the marriage flight, and a number of queens are
supported in each formicary. In some of the species the workers
are uniform in appearance, while in others they exhibit great differences in size and structure. As already stated, the workers or
neuters are generally divided into two classes, viz., the ordinary
small kind, and a second kind with much larger head and mandibles, and called soldiers. Bates has shown that in the Sauba
ant of South America (*Œcodoma cephalotes*) there are two forms
of the large-headed neuters, one with hairy and the other with
polished head.

LENGTH OF LIFE IN ANTS.—Lubbock's experiments have shown
that in some species the mature workers will live from one to six
years, and the females even much longer, the life of the males
being very ephemeral and lasting but a few days or weeks. He
kept a female of *Formica fusca* for thirteen years.

MIGRATIONS.—There are two kinds of ant migrations. The
swarming of the sexes takes place usually in the afternoon or toward evening on warm or sultry days, and it is remarkable
how very general, over a wide extent of country, the same
species will begin to fill the air on some particular day.
Species of the genera Lasius, Formica, Tetramorium, and Cremastogaster, particularly, often form dense swarms or clouds,
ascending high up into the air. These swarms of ants have
sometimes been known to be so dense and persistent that it was impossible, over large areas, to put the foot down without crushing
dozens of the insects which have been swept together in vast piles.
A case is on record of a large species covering the surface of the
water at sea to a depth of six inches, and for a distance of six
miles. This congregating in such vast swarms is due to the uniform and simultaneous hatching and development in all the colonies over a large extent of country.

The migrations of the sexes are really love excursions, whereas
the migrations of the workers, which take place in vast bodies at
times, are a result of undue multiplication, and are intended to
improve the condition of the surplus progeny and found new
colonies.

MYRMECOPHILÆ.—A most interesting lecture might be devoted

to the subject of myrmecophilous insects alone. Ants are as a
rule hostile to every other living thing, except such as the plant-
lice, which furnish them with desired sweets. They fiercely
resent any intrusion into their nests, and often attack and kill
their own kind if belonging to another colony. It is therefore remarkable that careful examination of almost any formicary will reveal the presence of a multitude of different insects which appear to live peaceably in the company of the legitimate inhabitants. A mere list of these myrmecophilous insects
would be of little interest. The species comprise, first, those
which, in the larva and pupa states, live among the ants; secondly,
accidental visitors, not confined to ants' nests; and, thirdly, the
true myrmecophilous species, i. e., those which in the imago state,
and so far as known in the adolescent states also, are exclusively
found in ants' nests and depend for their existence on the ants.
In some species of the second category we already find a tendency
to simulate in color the ant itself, or the surroundings of the
formicary; but the true myrmecophilæ, or species of the third
class, often mimic in the most remarkable manner the host upon
which they depend. Some of these myrmecophilous species are
mere scavengers, and feed upon the offal, of an animal or vegetal
nature, which is always found abundantly in the nests of ants.
They are endured with indifference by the ants, because they are
useful in an indirect way, helping in the performance of a duty
which would otherwise have to be performed by the ants themselves. Another group is present as marauders, living in the
nests for the purpose of stealing and devouring the ants' eggs,
larvæ or pupæ, whenever a chance offers. To this group belong the various Histeridæ, a Coleopterous family in which the
species are so constructed that it is impossible for the ants to ad-
vantageously attack them. In the third group we find species
characterized by sweet secretions, from which the ants derive
benefit. In some cases, as in the black, clumsy beetles of the
genus Cremastochilus, the insects are not absolutely confined to
the formicary, though they are always developed there. Frequently in the perfect state they endeavor to escape, and it is curious to note the strategy which the ants employ to prevent the de-
parture of these inquilines or guests from which they obtain the
coveted sweet. In such cases, as in the well known genus Claviger,
and allied genera, the insects are absolutely dependent on the ants,

which take the same tender care of them that they do of their
own young, feeding them and keeping them clean, and in every
way showing the utmost friendship.

TERMITES OR WHITE ANTS.

The Termites or White Ants have developed, in their higher
forms, an organization and a differentation of individuals very
similar to those of the true ants; whence the popular name.
They are among the oldest insects, as their remains are found in
the coal measures of Europe, whereas the true ants do not appear
until the Tertiary. Belonging, in fact, to an order which has
been very generally looked upon as the lowest or least developed
among the Hexapods and as representing most nearly the earlier
or primitive insects which appeared upon the globe, the fact that
they have acquired a social organization which in so many respects recalls that of the ants, is of great significance, as we shall
see when we come to consider the origin and development of these
traits. Yet a more intimate acquaintance with the facts concern-
ing the Termites shows us that the development of the social
habit and the differentation of forms, have been along different
lines from those presented by the social Hymenoptera, and are
based upon a different mode of development. In other words,
the Termites, belonging to an order which undergoes incomplete
metamorphoses—the larva being born in the image of the adult,
minus wings—is more or less capable of self-support soon after
birth, while in the social Hymenoptera, which undergo a complete
metamorphosis, the larva is quite unlike the adult, and entirely
helpless during development.

It is only within recent years that the Termites have been carefully studied. The results of these later studies must be rele-
gated to a note. (Note 6). While with most species the colony consists of a king and a queen and of two forms of neuters or workers;
yet in the European *Termes lucifugus* as many as fifteen distinct
forms have been characterized, but no true queen discovered. In
other words, besides the four distinctive classes of individuals
which characterize the more highly developed species, we find,
sometimes in the same species, but particularly when the different
species are considered, every gradation between these different
classes.

The fundamental difference between the social Hymenoptera

on the one hand, and the Termites on the other, is that in the
latter the workers or neuters (including the soldiers) are not undeveloped females, but consist of both sexes, and are in reality
arrested or modified larvæ, in which the sexual organs are but imperfectly developed or are completely atrophied. They are recognizable as neuters even after the first larval molt. The common
North American species, *Termes flavipes*, is doubtless familiar to
most of you. It occurs in vast numbers in rotten or prostrate logs,
and frequently invades our houses wherever there is wood in process of decay. The newly-hatched young are very tender and
helpless, and move but little, and while in the order Neuroptera
the young larva is usually able to care for itself immediately
after birth, the newly hatched Termite has become more or less
dependent upon the care of the workers, which either feed it with
partly digested food from their own mouths, or with their own
secretions, or else prepare food for it. The eggs are laid in large
numbers by fertile females or supplementary queens, but are carried long distances by the workers into chambers which are generally several feet underground, or else in the heart of otherwise
solid trees.

The queen in those species which normally possess one to each
colony, becomes helpless as she increases in size and gravity, for
she attains to many times the bulk of the ordinary neuters, which
are always unwinged. Winged males and females develop from
a special brood, and often in such numbers that in spring they
swarm until they literally fill the air. They are distinguished
from the rest by being more chitinized and darker in color. The
great majority of the swarming sexed individuals are doomed to
perish, either while on the wing or after falling to the ground,
for they are the favorite food of almost all other creatures. But
even where not devoured, most of them die without founding new
colonies. Swarming is not for the purpose of mating, but it is
to be looked upon as an incident in the excessive multiplication
of the species, and as a means of inducing cross-fertilization between different colonies.

Upon settling on the ground, the swarming individuals cast
off their wings, and if a couple of opposite sex are fortunate
enough to enter the outlying burrows of some colony already
founded, or to meet a few workers, they are capable of founding
a colony themselves. It is only after a female has been duly pro-

vided with a place of shelter or cavity that the mating really
takes place, from which time forth she becomes more or less
stationary and extremely fecund. She becomes, in short, a true
queen, and her escort remains with her and has been called a true
king; for here again the Termites differ radically from the social
Hymenoptera in that coition takes place repeatedly. There are,
however, supplemental queens or nymph queens, which seem to
be capable of laying eggs, probably parthenogenetically, and
which never develop their wings.

The great majority of the neuters are true workers, but a certain proportion of them, about one per cent., are so-called soldiers,
having enormously developed heads and powerful jaws, very
much as in the true soldier-ants, and fitted for no other purpose
than the defence of the colony. Both kinds of neuters are perfectly blind.

The habits and economy of our *Termes flaripes* may be looked
upon as typical of the family; but there are species in different
parts of the world in which (as in Calotermes) the workers, or (as
in Anoplotermes) the soldiers, are absent; others (as in Eutermes)
where the soldiers (nasuti) have a bill instead of jaws; others in
which the reproductive forms are reduced to the one royal pair;
and though the fact has not been absolutely observed, there are
probably Termites which produce only males and females, as
with ordinary insects, or as in allied families of the Neuroptera.
The accompanying diagram will very well illustrate the modes
of development and genealogy of the different forms in a typical
Termite colony, while some additional sexual forms of less certain
character or fixity, have been observed by Grassi in the European
Termes lucifugus, and called complementary kings and queens. In
those colonies which have no true royal pair, their place is taken
by supplementary royal pairs. (Note 6.)

Forms in a Termes Colony under normal Conditions.

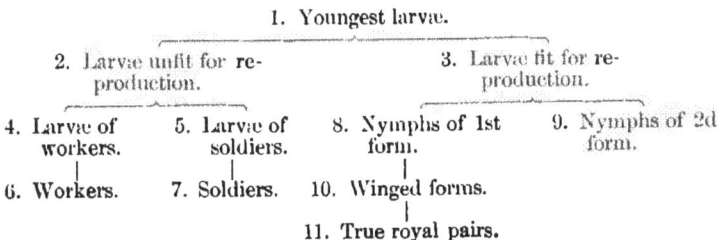

1. Youngest larvæ.

2. Larvæ unfit for re-production.	3. Larvæ fit for re-production.
4. Larvæ of workers. 5. Larvæ of soldiers.	8. Nymphs of 1st form. 9. Nymphs of 2d form.
6. Workers. 7. Soldiers.	10. Winged forms.
	11. True royal pairs.

The fecundity of the true queen Termite is something remark-able, and, based on Smeathman's observations on an African species (*Termes bellicosus*) the fact that an egg is produced every second, or some 80,000 a day in the height of the breeding season, has been commonly quoted among writers on the subject. In this species the queen is sealed up in a cell which is as hard as a stone, in the central and most protected part of the terni-tary, the cell being opened and enlarged from time to time by the workers, and being also perforated by holes which admit the workers to care for and feed her, while preventing the egress of the female and her attendant male escort.

Among the more curious facts connected with these Termites, because of their exceptional nature, is the late development of the internal sexual organs in the reproductive forms and the existence of a single long-lived male—a condition not parelleled among other insects, so far as I am aware. Further, as Dr. Hagen has pointed out, the queen represents a unique instance among insects of actual growth taking place in the imago state; for the intra-segmental ligaments not only expand, but grow with the increasing gravity of the abdomen, the stigmata actually taking part in this growth, though the dorsal abdominal plates remain unaffected.

In the Hive Bee multiplication of colonies takes place by divi-sion, but the colonizing swarm carries in itself all the elements necessary for the foundation of a new colony. In the more typi-cal Termites multiplication of colonies also takes place by division, but this is carried out by the neuters and the various adolescent stages, since there is usually but one true queen, which can not be moved. The new colony, therefore, can only obtain a true queen by introducing one of the royal pairs that wander about after they have swarmed and thrown off their wings. That great difficulty attends the establishment of such a royal pair of indi-viduals in a colony is illustrated by the fact that they are rarely discovered among colonies of our commoner species of Termes proper.*

---

*From the accounts of authors there is no difficulty in finding the true queen in most of the nest building species of Eutermes in the West Indies, Central and South America; while from Smeathman's famous account of *Termes bellicosus* in Africa, it would seem that the fertile queen is usually present in the colonies. But in the species most studied, viz., *Termes lucifu-*

The Termites thus exhibit a greater variety of resources for the perpetuation of the species, in case of emergency, than even the social Hymenoptera, and they also exhibit a greater variety of individual forms in the same colony. There is also among the different species, and especially among the different genera, a gradation from the simple to the more complex economy. Their habitations also vary from the simple to the more complete.

Calotermes burrows in the branches of trees and requires no specialized cells or chambers. *Termes flavipes* and allied species make extensive excavations in prostrate logs or the beams of houses, and are very destructive to old books, especially in dark and damp situations. The excavations are usually elongate and separated by partitions which are penetrated occasionally so as to connect the whole. The walls are lined with a thin layer of brown excrementitious matter, and some of the chambers are more particularly used to store eggs in, while others are used as nurseries for the young. Subterranean galleries often extend some distance away from the main termitary, and sometimes up under the bark of trees. More rarely they are exposed above ground, when the insects thicken the layer of excrementitious matter.

Eutermes, which is common in the West Indies and in Central and South America, builds exterior nests more or less spherical or conical, generally at the base of trees, but also on the branches or on stone walls. They are often as large as a hogshead, and consist chiefly of excrementitious matter and of collected particles of decayed wood. There are one or more queen cells in the most protected parts of the nest, and other chambers for the eggs and young, while temporary enlargements afford shelter for the winged individuals before swarming. Covered galleries somewhat thicker than an ordinary pencil, and composed of the same material as the nest, but less compact, extend from the main nest to the ground, or up the tallest trees, leading to food supplies.

The constructional faculty is yet more highly developed in the

---

gus, the difficulties in procuring a true queen would seem to be very great, and Prof. Grassi, in five years' observations, has never found one. Yet he had no difficulty in obtaining true kings and queens in confinement by establishing little colonies of winged individuals. The same condition of things prevails with our North American *Termes flavipes*, since in my own observations and those of others, no true queen has been met with, and reproduction is carried on, for the most part, by supplementary queens.

hill-making species of the genus Termes, which attain greatest perfection in South Africa. These nests always arise from the ground, and vary according to the species. They are made of finely comminuted wood, mixed with some secretion, or of clay, in which case they become as hard as stone. Long subterranean foraging galleries are extended from these nests.

In South America some species seem merely to excavate subterranean galleries in the soil, while Bates found at Santarem, Brazil, composite nests occupied by different species, which built each its own part of the nest with its own special material.

Some Generalizations.

In the hasty summary which I have thus endeavored to present to you of some of the chief characteristics of social insects, those who are most familiar with the facts can best appreciate how much of interest has been omitted. These insects are attacked by various natural enemies in their own class, and particularly in the case of the bees and wasps, by some of the most abnormal parasites, viz., the Stylopidæ, in which the young larva is extremely active, but the adult female stationary and so degraded that she has lost all members and mouth-parts, and in fact all semblance of an insect, while the adult male is an active, winged creature, of very ephemeral existence. Chapters might be written upon the myrmecophilous and termitophilous insects of various orders, some of which are mere mess-mates, others advantageous associates, while others are unwelcome, but more or less successful intruders on the hospitality of their hosts. This part of the subject must, however, be passed over in order to permit me to close with some generalizations and speculations which the facts already enumerated provoke.

The Senses in Insects.

Having thus dealt, in a summary way, with some of the structures and economics of the social insects, let us now consider their psychological manifestations.

Of the five ordinary senses recognized in ourselves and most higher animals, insects have, beyond all doubt, the sense of sight, and there can be as little question that they possess the senses of touch, taste, smell and hearing. Yet, save, perhaps, that of touch, none of these senses, as possessed by insects, can be strictly compared with our own, while there is the best of evidence that insects possess other senses which we do not, and that they have sense organs with which we have none to compare. He who tries to comprehend the mechanism of our own senses—the manner in which the subtler sensations are conveyed to the brain—will realize how little we know thereof after all that has been written. It is not to be wondered at, therefore, that authors should differ as to the nature of many of the sense organs of insects, or that there should be little or no absolute knowledge of the manner in which the senses act upon them. The solution of psychical problems may never, indeed, be obtained, so infinitely minute are the ultimate atoms of matter; and those who have given most at-

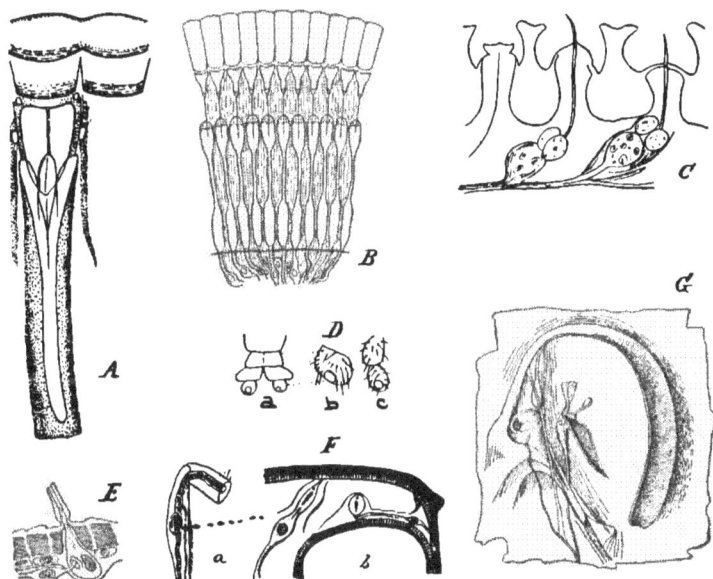

FIG. 8.—SENSORY ORGANS IN INSECTS: *A*, one element of the eye of Cockroach (after Grenacher); *B*, diagrammatic section of compound eye in insect (after Miall & Denny); *C*, organs of smell in Melolontha (after Kraepelin); *D, a, b,* sense organs of abdominal appendages of Chrysopila, *c*, small pit on terminal joint of palpus in Perla (after Packard); *E*, diagram of sensory ear of insect (after Miall & Denny); *F*, auditory apparatus of Meconema, *a*, fore tibia of this locust, *b*, diagrammatic section through same (after Graber); *G*, auditory apparatus of Caloptenus seen from inner side, showing tympanum, auditory nerve, terminal ganglion, stigma and opening and closing muscle of same, as well as muscle of tympanum membrana (after Graber).—All very greatly enlarged.

tention to the subject must echo the sentiment of Lubbock, that the principle impression which the more recent works on the intelligence and senses of animals leave on the mind, is, that we know very little indeed on the subject. We can but empirically observe and experiment, and draw conclusions from well attested results.

SIGHT.—Taking first the sense of sight, much has been written as to the picture which the compound eye of insects produces upon the brain or upon the nerve centers. Most insects which undergo complete metamorphoses possess in their adolescent states simple eyes or ocelli, and sometimes groups of them of varying size and in varying situations. It is difficult, if not impossible, to demonstrate experimentally their efficiency as organs of sight ; the probabilities are that they give but the faintest impressions, but otherwise act as do our own. The fact that they are possessed only by larvæ which are exposed more or less fully to the light, while those larvæ which are endophytous, or otherwise hidden from light, generally lack them, is in itself proof that they perform the ordinary functions of sight, however low in degree. In the imago state the great majority of insects have their simple eyes in addition to the compound eyes. In many cases, however, the former are more or less covered with vestiture, which is another evidence that their function is of a low order, and lends weight to the view that they are useful chiefly for near vision and in dark places. The compound eyes are prominent and adjustable in proportion as they are of service to the species, as witness those of the common House-fly and of the Libellulidæ or Dragon-flies. It is obvious from the structure of these compound eyes that impressions through them must be very different from those received through our own, and, in point of fact, the late experimental researches of Hickson, Plateau, Tocke and Lemmermann, Pankrath, Exner and Viallanes, practically establish the fact that while insects are short-sighted and perceive stationary objects imperfectly, yet their compound eyes are better fitted than the vertebrate eye for apprehending objects set in relief or in motion, and are likewise keenly sensitive to color.

So far as experiments have gone, they show that insects have a keen color sense, though here again their sensations of color are different from those produced upon us. Thus, as Lubbock has shown, ants are very sensitive to the ultra violet rays of the spectrum, which we cannot perceive, though he was led to conclude that to the ant the general aspect of nature is presented in an aspect very different from that in which it appears to us. In reference to bees, the experiments of the same author prove clearly that they have this sense of color highly developed, as indeed might be expected when we consider the part they have played in the development of flowers. While these experiments seem to show that blue is the bees' favorite color, this does not accord with Albert Müller's experience in nature, nor with the general experience of apiarians, who, if asked, would very generally agree that bees show a preference for white flowers.

TOUCH.—The sense of touch is supposed to reside chiefly in the antennæ or feelers, though it requires but the simplest observation to show that with soft-bodied insects the sense resides in any portion of the body, very much as it does in other animals. In short, this is the one sense which, in its manifestations, may be conceded to resemble our own. Yet it is evidently more specialized in the maxillary and labial palpi and the tongue than in the antennæ, in most insects.

TASTE.—Very little can be positively proved as to the sense of taste in insects. Its existence may be confidently predicated from the acute discrimination which most monophagous species exercise in the choice of their food, and its location may be assumed to be the mouth or some of the special trophial organs which have no counterpart among vertebrates. Indeed certain pits in the epipharynx of many mandibulate insects, and, in the ligula and the maxillæ of bees and wasps are conceded, by the authorities, to be gustatory.

SMELL.—That insects possess the power of smell is a matter of common observation, and has been experimentally proved. The many experiments of Lubbock upon ants left no doubt in his mind that the sense of smell is highly developed in them. Indeed it is the acuteness of the sense of smell which attracts many insects so unerringly to given objects, and which has led many persons to believe them sharp-sighted. Moreover, the innumerable glands and special organs for secreting odors, furnish the strongest indirect proof of the same fact. Some of these, of which the osmaterium in Papilionid larvæ and the eversible glands in Parorgyia are conspicuous examples, are intended for protection against inimical insects or other animals; while others, possessed by one only of the sexes, are obviously intended to please or attract. A notable development of this kind is seen in the large gland on the hind legs of the males of some species of Hepialus, the gland being a modification of the tibia, and sometimes involving the abortion of the tarsus, as in the European *H. hectus* L. and our own *H. behrensi* Stretch. The possession of odoriferous glands, in other words, implies the possession of olfactory organs. Yet there is among insects no one specialized olfactory organ as among vertebrates; for while there is conclusive proof that this sense rests in the antennæ with many insects, especially among Lepidoptera, there is good evidence that in some Hymenoptera it is localized in an ampulla at the base of the tongue, while Graber gives reasons for believing that in certain Orthoptera (Blattidæ) it is located in the anal cerci, and the palpi.

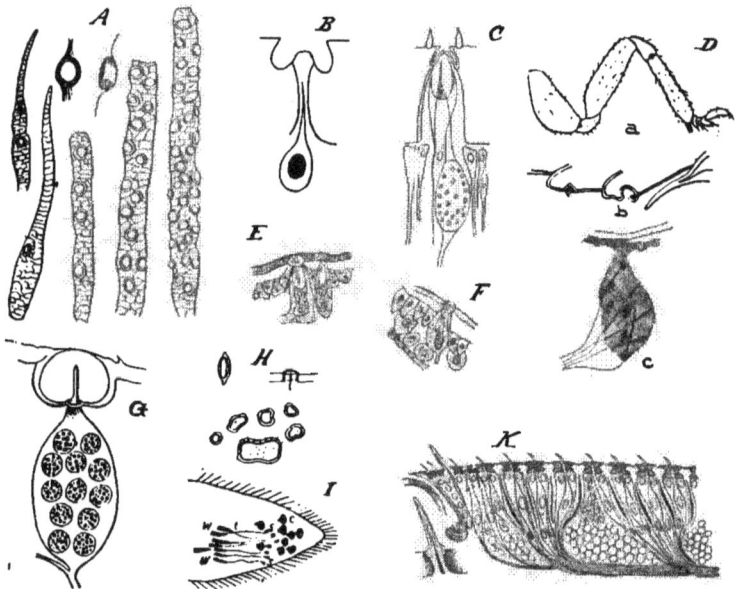

FIG. 9.—SENSORY ORGANS IN INSECTS : *A*, sensory pits on antennæ of young wing less *Aphis persicæ-niger* (after Smith); *B*, organ of smell in May Beetle (after Hauser); *C*, organ of smell in Vespa (after Hauser) ; *D*, sensory organs of *Termes flavipes*, *a*, tibial auditory organ, *c*, enlargement of same, *b*, sensory pits of tarsus (after Stokes); *E*, organ of taste in maxillæ of *Vespa vulgaris* (after Will) ; *F*, organ of taste in labium of same insect (after Will); *G*, organ of smell in Caloptenus (after Hauser); *H*, sensory pilose depressions on tibia of Termes (after Stokes) ; *I*, terminal portion of antennæ of *Myrmica ruginodis*, *c*, cork shaped organs, *s*, outer sac, *t*, tube, *w*, posterior chamber (after Lubbock) ; *K*, longitudinal section through portion of flagellum of antennæ of worker bee, showing sensory hairs and supposed olfactory organs (after Cheshire).—All very greatly enlarged.

HEARING.—In regard to the sense of hearing the most casual experimentation will show (and general experience confirms it) that most insects, while keenly alive to the slightest movements

or vibrations, are for the most part deaf to sounds which affect
us. That they have a sense of sound is equally certain, but its
range is very different from ours. A sensitive flame arranged for
Lubbock by the late Prof. Tyndall, gave no response from ants,
and a sensitive microphone arranged for him by Prof. Bell gave
record of no other sound than the patter of feet in walking. But
the most sensitive tests we can experimentally apply may be, and
doubtless are, too gross to adjust themselves to the finer sensibil-
ities of such minute, active and nervous creatures. There can
be no question that insects not only produce sounds, but receive
the impression of sounds entirely beyond our own range of per-

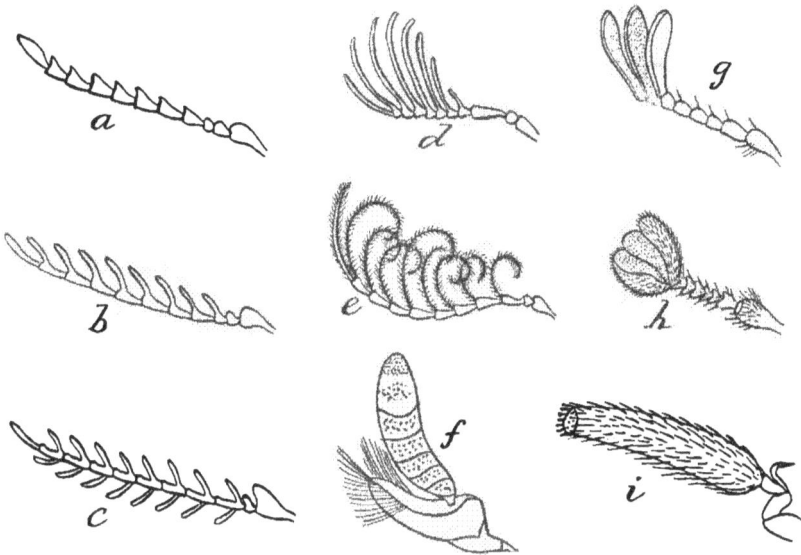

FIG. 10.—SOME ANTENNÆ OF COLEOPTERA : *a*, Ludius ; *b*, Corymbites ; *c*, Prionocy-
phon ; *d*, Acneus ; *e*, Dendroides ; *f*, Dineutes ; *g*, Lachnosterna ; *h*, Bolbocerus ; *i*,
Adranes, (after LeConte and Horn).—All greatly enlarged.

ception, or as Lubbock puts it, that " we can no more form an
idea of than we should have been able to conceive red or green
if the human race had been blind. The human ear is sensitive
to vibrations reaching at the outside to 38,000 in a second. The
sensation of red is produced when 470 millions of millions of
vibrations enter the eye in a similar time ; but between these two
numbers, vibrations produce on us only the sensation of heat. We
have no especial organ of sense adapted to them." It is quite certain that ants do make sounds, and the sound-producing organs on

some of the abdominal joints have been carefully described. The
fact that so many insects have the power of producing sounds
that are even audible to us, is the best evidence that they possess
auditory organs. These are, however, never vocal, but are situ-
ated upon various parts of the body or upon different members
thereof.

SPECIAL SENSES AND SENSE-ORGANS.—While from what has
preceded it is somewhat difficult to compare the more obvious
senses possessed by insects with our own, except, perhaps, the
sense of touch, it is, I repeat, just as obvious to the careful
student of insect life that they possess special senses which it is
difficult for us to comprehend. The sense of direction, for instance, is very marked in the social Hymenoptera which we have
been considercring, and in this respect insects remind us of many
of the lower vertebrates which have this sense much more strongly
developed than we have. Indeed, they manifest more especially
what has been referred to in man as a sixth sense, viz., a certain
intuition which is essentially psychical, and which undoubtedly
serves and acts to the advantage of the species as fully, perhaps,
as any of the other senses. Lubbock demonstrated that an ant
will recognize one of its own colony from among the individuals
of another colony of the same species, and when we consider that
the members of a colony number at times not thousands but
hundreds of thousands, this remarkable power will be fully appreciated.

The neuter Termites are blind and can have no sense of light
in their internal or subterranean burrowings; yet they will
undermine build-
ings and pulverize
various parts of
elaborate furni-
ture without once
gnawing through
to the surface,
and those species
which use clay
will fill up their
burrowings to

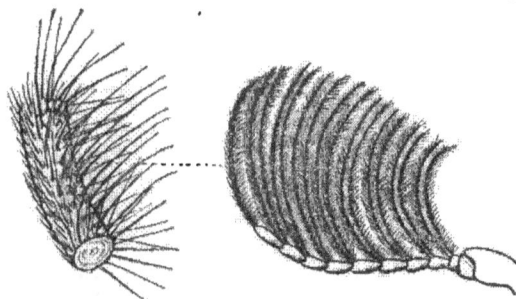

FIG. 11.—Antenna of male Phengodes with portion of ray. —Greatly enlarged. (Original.)

strengthen the supports of structures which might otherwise fall
and injure the insects or betray their work. The bat in a lighted

room, though blinded as to sight, will fly in all directions with such swiftness and with such infallible certainty of avoiding concussion or contact, that its *feeling at a distance* is practically incomprehensible to us.

The manner in which anything threatening its welfare thrills and agitates one of these insect communities, and causes every individual to act at once for the common good, has been noted by all observers, and is a good illustration in point. It may be likened to the manner in which the same conditions influence communities of other animals, including man. There are emergencies when intuitive feeling dispossesses reason, and every capable person seems blindly urged to definite action for the protection of the community, regardless of consequence. The war-cry of a nation is an example in point, and violations of otherwise just, but tedious, processes of law, are under certain circumstances deemed justifiable. I shall never forget the emotion that influenced the citizens of Chicago the day following their great fire in 1871. Reason, argument, judgment, were in abeyance. The quicker, intuitive processes prevailed, and to meet lawlessness and the tendency to incendiarism, every right-minded citizen was ready to do vigilant duty, regardless of personal interest, every incendiary being hung to the nearest lamp-post, without ado or delay. It was the universal and deep-seated instinct of self-preservation.

TELEPATHY.—But however difficult it may be to define this intuitive sense which, while apparently combining some of the other senses, has many attributes peculiar to itself, and however difficult it may be for us to analyze the remarkable sense of direction, there can be no doubt that many insects possess the power of communicating at a distance, of which we can form some conception by what is known as telepathy in man. This power would seem to depend neither upon scent nor upon hearing, in the ordinary understanding of these senses, but rather on certain subtle vibrations, as difficult for us to apprehend as is the exact nature of electricity. The fact that man can telegraphically transmit sound almost instantaneously around the globe, and that his very speech may be telephonically transmitted, as quickly as uttered, for thousands of miles, may suggest something of this subtle power, even though it furnish no explanation thereof.

The power of sembling among certain moths, for instance, es-

pecially those of the family Bombycidæ, is well known to ento-
mologists, and many remarkable instances are recorded. (Note 7.)
I am tempted to put on record, for the first time, an individual
experience which very well illustrates this power, as, on a number of occasions when I have narrated it, most persons not familiar
with the general facts have deemed it remarkable. In 1863 I obtained from the then Commissioner of Agriculture, Col. Capron,
eggs of *Samia cynthia*, the Ailanthus silk worm of Japan, which
had been recently introduced by him. I was living on East

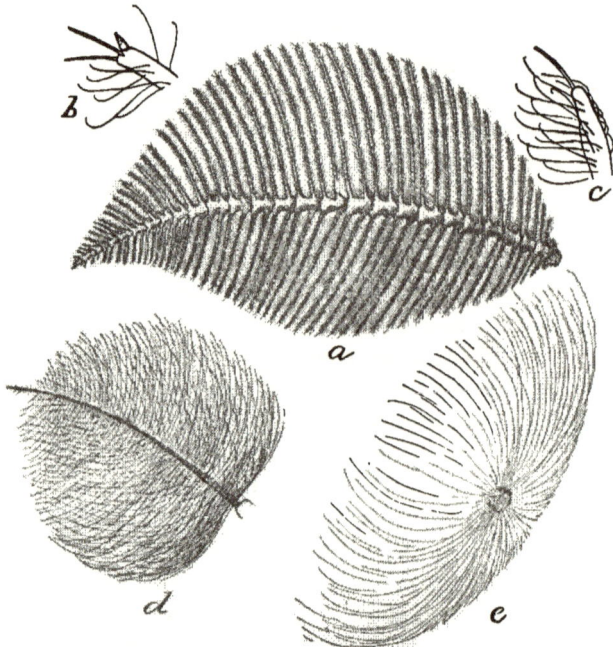

FIG. 12.—SOME ANTENNÆ OF INSECTS: *a*, Telea polyphemus, male, x 3 ; *b* and *c*, tips
of rays of same still more enlarged ; *d*, Chironomus x 6 ; *e*, section of same still more
enlarged. (Original.)

Madison Street, in Chicago, at the time, a part of the city subsequently swept by the great fire, and since entirely transformed.
In the front yard, which (so commouly the case in the old
Chicago days) was below the sidewalk, there grew two Ailanthus
tress which were the cause of my sending for the aforesaid eggs.
I had every reason to believe that there were no other eggs of
this species received in any part of the country within hundreds
of miles around. It seemed a good opportunity to test this power

of sembling, and after rearing a number of larvæ, I carefully watched for the appearance of the first moths from the cocoons. I kept the first moths separate and confined a virgin female in an improvised wicker cage out of doors on one of the Ailanthus trees. On the same evening I took a male to the old Catholic cemetery on the north side, which is now a part of Lincoln Park, and let him loose, having previously tied a silk thread around the base of the abdomen to insure identification. The distance between the captive female and the released male was at least a mile and a half, and yet the next morning these two individuals were together.

Now in the moths of this family the male antennæ are elaborately pectinate, the pectinations broad and each branch minutely hairy. (See Fig. 12, *a*). These feelers vibrate incessantly, while in the female, in which the feelers are less complex, there is a similar movement connected with an intense vibration of the whole body and of the wings. There is therefore, every reason to believe that the sense is in some way a vibratory sense, as indeed at base is true of all senses, and no one can study the wonderfully diversified structure of the antennæ in insects, especially in males, as very well exemplified in some of the commoner gnats (see Fig. 12, *d*, *c*), without feeling that they have been developed in obedience to, and as a result of, some such subtle and intuitive power as this of telepathy. Every minute ramification of the wonderfully delicate feelers of the male Mosquito, in all probability pulsates in response to the piping sounds which the female is known to produce, and doubtless through considerable distance.

There is every justification for believing that all the subtle cosmic forces involved in the generation and development of the highest, are equally involved in the production and building up of the lowest of organisms, and that the complexing and compounding and specialization of parts have gone on in every possible and conceivable direction according to the species. The highly developed and delicate antennæ in the male Chironomus, for instance, may be likened to an external brain, its ramifying fibers corresponding to the highly complicated processes that ramify from the nerve cells in the internal brains of higher animals, and responding in a somewhat similar way to external impressions. While having no sort of sympathy with the foolish notions that the spiritists proclaim, to edify or terrify the gullible and unscientific, I am just as much out of sympathy with that class of materialistic scientists who refuse to recognize that there may be and are subtle psychical phenomena beyond the reach of present experimental methods. The one class too readily assumes supernatural power to explain abnormal phenomena; the other denies the abnormal because it likewise is past our limited understanding. "Even now," says William Crookes, who speaks with authority, "telegraphing without wires is possible within a radius of a few hundred yards," and, in a most interesting contribution to our present knowledge of vibratory motion and the possibilities of electricity, the same writer remarks :*

"The discovery of a receiver sensitive to one set of wave lengths and silent to others is even now partially accomplished. The human eye is an instance supplied by nature of one which responds to the narrow range of electro-magnetic impulses between the three ten-millionths of a millimeter and the eight ten-millionths of a millimeter. It is not improbable that other sentient beings have organs of sense which do not respond to some or to any of the rays to which our eyes are sensitive, but are able to appreciate other vibrations to which we are blind. Such beings would practically be living in a different world from our own. Imagine, for instance, what idea we should form of surrounding objects were we endowed with eyes not sensitive to the ordinary rays of light, but sensitive to the vibrations concerned in electric and magnetic phenomena. Glass and crystal would be among the most opaque of bodies. Metals would be more or less transparent, and a telegraph wire through the air would look like a long narrow hole drilled through an impervious solid body. A dynamo in active work would resemble a conflagration, while a permanent magnet would realize the dreams of mediæval mystics and become an everlasting lamp with no expenditure of energy or consumption of fuel.

In some parts of the human brain may lurk an organ capable of transmitting and receiving other electrical rays of wave lengths hitherto undetected by instrumental means. These may be instrumental in transmitting thought from one brain to another." * * *

INTELLIGENCE IN INSECTS.

Anyone who has closely studied the ways of insects, especially as exemplified in the social species we have been considering, will not doubt that they possess intelligence. They communicate with each other by a language, which, though unspoken, is no less eloquent of all their wishes and desires. They work for the

---

*"Some possibilities of electricity." *Fortnightly Review*, March, 1892.

common good; they train a soldier and a police force; they are brave in the defense of the communal interest; they protect and defend their sovereign; they make war and even organize military expeditions; they make slaves and are held in bondage; they encourage and protect other insects which yield them cherished nourishment; they even go so far as to care for the eggs of such, and thus deliberately rear their nectar-giving kine; they cultivate crops; they providently store food for winter use or in anticipation of an inauspicious season; they give expression to satisfaction and pleasure; they exhibit certain baser passions, as jealousy, ill-temper or rage; they even display a certain moral sense, and will help, on occasions, the distressed and threatened of their own kind; they are most assiduous in the care and rearing of their young; they profit by experience; they manifest a pure and simple enjoyment of life by their gambols and playfulness; they are cleanly to a degree which will astonish those who for the first time observe their constant licking and brushing of all parts of the body; they exhibit, in short, most of the sense manifestations displayed by higher animals. It may be ingeniously argued that in all these manifestations they are acting as mere automatons, but the same arguments may be, and have been, urged to explain the actions of man.

So far as experimentation goes, and especially that by Sir John Lubbock, bees are not gifted with the high degree of intelligence with which many writers have credited them, and in this respect do not compare with the higher ants. The Termites are probably the lowest, bees next, wasps next, and ants the highest, in point of intelligence, among social insects. The affection of the bees for their queen, or the deference paid by ants, wasps and Termites to theirs, may be viewed as an instinctive expression of their communal obligation to her, which is at once transferred to another by whom she may be replaced; but our own fealty to our rulers may bear very much the same interpretation. Wasps are more alert and intelligent than bees, and as Lubbock has shown, are measurably susceptible of being tamed. Ants, as we have seen, exhibit a very high degree of intelligence. In fact, the manner in which all these insects work together in harmony, and especially the manner in which certain individuals act as scouts or deliberately set to work to remedy and overcome any exceptional interference with or injury to their habitations, denotes consider-

able reasoning and reflective power ; while the anticipation of the
needs of the community—as in the building of queen cells at the
proper time by bees and wasps, the varying treatment of the
young, and the preparation for swarming by all the social insects
—argues an intuitive perception which is as conscious as that
which higher animals display in making similar provision for
their progeny. This very intuition is the origin of intellect, or,
rather, the primary form of intellect. It is, as Ward expresses it,
older than reason, and parent of the later faculties of abstraction
and reflection. It involves all that we know as sagacity and cun-
ning, displayed by animals for their own good.

But it is to the nature of this intelligence that I would call
your attention, since many may question the use of the term in
connection with these insects. It has been the fashion in the
past to separate man from the rest of the animal world by the
nature of his intelligence. The earliest philosophers, instead of
beginning with the simpler problems of subjective nature, seem
to have been fascinated by the more complex phenomena of ob-
jective nature. They built up a fabric of metaphysics which
modern methods of induction and modern experimental physi-
ology and psychology have demolished and remodeled. We have
had dissertations on the will as something quite independent of
the body, and speculations as to the difference between human
and divine will.

We must certainly grant to insects the sensations of pleasure
and pain, for the worthiest authorities now concede that the
least of sentient beings—or animals as contra-distinguished from
plants—must possess feeling, however faint. Feeling means
either pleasure or pain, the former the inevitable out-growth of
experience favorable to the organism, the latter the converse.
The former is a sign post on the road to all that is good for the
race, the latter a warning of all that is evil; though, para-
doxical as it may seem, this is just as necessary to the welfare of the organism. What is evil for the individual may be
good for the race. Now all feeling must be conscious, and the
different grades of consciousness of feeling, until we reach self-
consciousness, involving intellectual processes, are but gradations
in the manifestations of one and the same kind of force. Indeed,
it is now conceded by advanced thinkers of the biologic school
that intellect had its origin in and depends on the senses, and

that mind is divisible into feeling and understanding. Most of the acts of these social insects are, it is true, what we call instinctive*; but as I have so often had occasion to express my views and the reasons for them, on the subject of instinct, it is unnecessary to enlarge upon them further than to state that the instinctive acts of insects are often combined, in a greater or less degree, with a low order of conscious reasoning, and that while this is generally of the intuitive kind, it is, on occasions, deliberate and reflecting.

> "If in the insect Reason's twilight ray
> Sheds on the darkling mind a doubtful day,
> Plain is the steady light her Instincts yield,
> To point the road o'er life's unvaried field ;
> If few those instincts, to the destined goal,
> With surer course, their straiten'd currents roll."—Evans.

Two beliefs that have very generally prevailed among men up

---

*Romanes considers that the instincts of neuter insects are themselves sufficient to refute Lewes' theory of instinct as being lapsed intelligence transmitted through heredity ; and he criticises Spencer's views that "the automatic actions of a bee building one of its wax cells answer to outer relations so constantly experienced that they are, as it were, organically remembered." He bases his criticism upon the statement that the bee "begins by performing these actions before it has itself had any individual experience of cell-making and without its parents ever having had any ancestral experience." While this statement represents accepted belief, it follows from what I have already said of the bee that it is essentially untrue. The worker could no more begin to secrete wax and build cells until it had acquired a certain age than could mammals secret the lacteal fluid before a certain age ; and during its early life as an adult it had the experience of its older fellows to guide it, were such guidance necessary. The example chosen by Romanes was simply unfortunate. To understand the development of the cell-building instinct, we must consider the stages of its development as illustrated in the varying forms of cells yet existing, from the cruder cells of Bombus on, and remember that each step in the more perfect building has been accompanied by structural modifications, and that the instincts have been accumulated and perfected by heredity *pari passu* with the structures ; further that the habit probably became so firmly fixed before the neuters had been differentiated, that it has been transmitted since that time through the queen, though she herself no longer possesses it ; further that while instinctive performance is ordinarily inevitable, it yet varies in the amount of its fixity and accuracy and often leads astray or fails ; and, finally, that it is often modified by individual experience or reason, or by communal interest or necessity—these truths applying particularly to the social insects, and in a variable degree to all animals.

to within recent years, have been so effectually discarded that they are even renounced by the more advanced theologians. I refer to the belief that organisms were specially created as they now exist, and that man was apart from, and not a part of, the rest of the animal world. It is my judgment that a third equally prevalent notion is essentially false, and will have to be abandoned before we can properly appreciate the psychology of animals. I refer to the notion that the lower animals do not reason, and are incapable of conscious reflection and thought. It would be easy to occupy your time for hours with accounts of their actions which can be explained only upon the views here set forth, and which are utterly at variance with the popular notions and prejudices.

The insects to which I have referred to-night are admitted to be among the more intelligent of their class; but they are only illustrations of an intelligence which is found throughout the other orders, and which impresses us in proportion as we study it and come to realize and recognize it. We can never properly appreciate, nor properly bring ourselves into sympathy with these lower creatures, until we recognize that they are actuated by the same kind of intelligence as we ourselves. There are certain acts which all creatures necessarily perform, as an outgrowth of their organization. These are essentially the instinctive acts, and are, for the most part, inevitable and often unconscious. A great many of the acts of rational men are, in this view, instinctive, and from birth to maturity many of them are prompted solely by the consecutive development of different parts of the organization, and are much less the result of training and teaching than is generally believed. Most of the acts of insects are instinctive and explicable upon this same view, but no one can study them carefully and without bias and not feel that these instinctive and inevitable actions are associated with many others which result from the possession of intelligence—of conscious reasoning and reflective powers. In this view of the case is the whole world truly kin, and is man brought more fully into sympathy with and appreciation of it.

Is it not significant also, that, just as in man, among mammalia, the higher intellectual development and social organization is found correlated with the longest period of dependent infancy; that this helpless infancy has been, in fact, as Fiske has shown,

a prime influence in the origin, through family, clan, tribe and
state, of organized civilization; so in the insect world we find
the same correlation between the highest intelligence and dependent infancy, and are justified in concluding that the latter is, in
the social Hymenoptera as in man, in the same way the cause
of the high organization, and division of labor so charateristic
of them!

HEREDITY: NATURAL SELECTION.

The application of the principle of natural selection to the production of neuter insects, and especially to the production of
neuter insects of diversified form, seems, at first sight, impossible. Indeed, we know that Darwin felt that this question of
neuter insects was one of the most difficult to deal with in connection with his grand generalization. Weismann, who believes
in the all-sufficiency of natural selection, insists, and has within
the past year, in his controversies with Herbert Spencer, emphasized his belief, that these neuter insects absolutely preclude the
idea of the transmission of acquired characters, and endeavors to
explain their occurance by his own peculiar theories as to modification taking place in the germ plasm. I shall certainly not
attempt, in the limited time that I may yet hope to hold your
attention, to discuss in detail the views held whether by Weismann or his opponents;* but I will venture to show that, while

---

*The chief argument in favor of Weismann's theory of heredity is that
it is an earnest attempt to establish a basis in observed histologic and em-
bryologic facts. The idea of the continuity or "immortality" (using the
word in his own qualified use of it) of the germ plasm is a bold one which
gives us at least a conceivable and material basis of reproduction, and is
justified, though not absolutely, in the facts referred to and in the history
of the Protozoa. One of the chief arguments against it is, in my judgment,
that, inasmuch as it precludes the transmission of impressions on the soma, i.
e., individually acquired characters, Weismann has, in order to sustain the
theory, been led to question and finally to deny the transmissibility of such acquired characters. It is difficult to formulate the later modifications of the original theory without using many Weismannisms, themselves requiring chapters
of explanation; but that variation is due to direct effects on the germ plasm
by inequalities of nutrition, is, I believe, a correct statement of his latest
views. The trouble with all theories of reproduction and heredity based
solely on observed microscopic facts, is that the essence, the life principle,
the potential factors, must always escape such methods. Wherefore any
theory that will hold must cover the psychical as well as the physical facts
—the total of well established experience—and this truth was doubtless

the social insects offer the most serious obstacles to the accept-
ance of the theory of natural selection as an all-sufficient theory
to explain the phenomena, yet the facts are perfectly explicable
upon the général principles that have governed the modification
of organisms, among which that of natural selection plays an important, but limited part.

In the economy of the Hive Bee we have seen that all the
neuters are structurally alike, and that the different functions
which they perform result from inherited tendencies or structural
peculiarities developed at different ages. There are some records
of abnormal workers, small drones, and slight variations in the
amount of arrestation of development; but on the whole the
three classes of queen, worker and drone are remarkably well
differentiated and fixed. We have seen that the differences in the
two former classes result from conditions of food, treatment and
environment of the young, and are under the control of the
colony. Each fertile egg has the potentiality of developing a
fertile queen, and as the neuters, under exceptional conditions,
are able to lay eggs which invariably produce drones, the queen,
through such drones, must occasionally inherit indirectly from
the workers. At bottom, however, the differentiation between
the workers and the queen is purely a matter of food and bringing up, or *education*, as the French would more correctly call it.
In other words, the ultimate result is decided for each generation
in the treatment of the young or the larvæ. The drone results
from an unfertilized egg, and as the egg is only fertilized when
the tip of the queen's abdomen is pressed into a worker cell, and
not when thrust into a drone cell, the production of drones is
also under control of the colony.

I have already called attention to the fact that other species

recognized by Darwin in framing his tentative theory of pangenesis.

Weismann's efforts to derive a physical theory of reproduction and evolution find a paralell in the efforts of those entomological histologists who,
starting with the conception that the development of the individual was
but an unfolding of structures already nascent in the embryo, expected to
find—and even claim to have found—all the structures of the imago represented, *in petit*, in the larva. In truth, however, there is a total re-adjustment
of cells, and development *de novo* of organs, with each important change or
molt, and the vital force which impels this development, whether of the
minutest bodily structure or the subtlest intellectual attribute, is the great
mystery beyond explanation.

of bees show gradations in these two kinds of females, and that some species permit more than one queen or fertile female in the colony and would refer for further details, both as to present gradations and variations to the Notes, especially numbers 1, 2, 3, and 4. Natural selection, if it has played any part at all, must have done so chiefly in the manner ingeniously suggested by Darwin himself, namely, not as between individuals, but as between colonies. The tendency to produce arrested females or neuters doubtless became fixed in some ancestral form through social selection, and is kept up by this and colony selection.

In the wasps we have a very different state of things, involving the parthenogenetic production of arrested females and the seasonal production of fully developed forms of both sexes. Here again, the evidence all goes to show that the differences depend for each generation on the environment, food and method of nurture of the larva, the tendency having become fixed in varying degrees in the different species, and only so fixed by being transmitted through the queen or sexually perfect females. So far as natural selection has acted at all, it has acted on the potentiality or inherited tendencies of these females. Very exact information is not yet at hand as to how far the neuters are variable, whether as to condition of the reproductive organs or as to size. But judging merely by mounted specimens which I have examined in various species, it is probable that there is some variation in these respects, though the three classes are quite neatly differentiated, much as in the bees.

When it comes to the ants, the problem is more complicated; but we may safely assume that the different forms have been brought about by the same influences. In a large colony of individuals, where size and character are not fixed by a definite cradle, but where the young larvæ are free and are carried about, nursed and fed by the workers, there would naturally arise greater variations between individuals, and while the kind of nourishment, or the kind of nurture, or the age of the female at the time the ova are produced, or the season of the year, have doubtless all contributed to the variation, and may still independently contribute to it at the present time; yet, whatever the causes of this variation, it has become fixed in certain definite lines that are more or less useful to the species. Whether or not the proportion of the different individuals is under the control of the colony as a whole, by virtue of the treatment of the larva, it will always be difficult to prove, though there is every reason to believe that, as in the bees, there is, to some extent, such control, and that the relative proportions of the different forms will depend upon circumstances. But the fact remains that, in ants, as in bees and wasps, the neuters are but arrested females, and are capable of becoming, under exceptional circumstances, fertile, and that we see in the different species all gradations, not only as to the number of forms of the workers, but as to the number of fertile females that are allowed in the same colony to provide for the continuance of the species. We also find in the same species great variation and gradation in the characters of the different sets which form the community, especially between the different forms of workers, in contrast to what I have remarked as to bees and wasps. This has been recorded not only by writers like Darwin and Lubbock, but by all who have given close attention to the subject; while Ch. Lespès (*Ann. des Sciences Nat.* (4) 20, pp. 241–251) in his "Observations sur les Fourmis Neutres" has shown that all neuters have traces of the female reproductive organs; that these traces vary in the different species; and that where there are two forms of neuters these pass insensibly into each other through intermediate forms. The ants thus furnish us with varying degrees of social organization when the different species are considered, while the different classes in the same species are not as definitely fixed as in the bees or the wasps.

Now it were comparatively easy to account for these neuters among the social Hymenoptera and the different forms and attributes which they present, by putting aside natural selection, as expounded by Darwin, and substituting therefor social selection acting not on generations in time, but on the individual at once by the manner of its bringing up; and surely there would seem to be sufficient justification for this course when we find not only such great physiological and functional, but such profound structural modifications induced by larval environment and nurture, as I have pointed out, especially between the queen and the worker bee.*   This has, in fact, been the chief explanation which

*Mr. Herbert Spencer, in one of his rejoiners to Prof. Weismann, (*Contemporary Review*, December, 1893) refers to a chapter on The Determination of Sex by Prof. Geddes and Mr. Thompson in their "Evolution of Sex,"

I have offered for the facts, in discussions with friends and before the society, limiting the action of natural selection to colonies
as a whole. Few persons who have not had large experience in
rearing insects can appreciate the full influence of larval environment and food on the ultimate imago, or the power of larval
accomodation to various conditions. All insects in the larva
state possess this power, within varying limits, and it is nowhere
more marked than in the Aculeate Hymenoptera. I have called
attention to it on numerous occasions* when treating of parasitic
species, and it is particularly noticeable in the fossorial Hymenoptera and the Meloïdæ. Size, especially, may easily be dimin-

where they state that "such conditions as deficient or abnormal food," and
others "causing preponderances of waste over repair * * * tend to result in the production of males," while "abundant and rich nutrition" and
other conditions which "favor constructive processes * * * result in
the production of females." He then cites J. H. Fabre's statement that in
the nests of *Osmia tricornis* the eggs at the bottom of the cell which are
first laid and accompanied by much food, produce females, while those at
the top, laid last and accompanied by one-half or one-third the quantity of
food, produce males. (Souvenirs Entomologiques, 3ème série, page 328).
He further refers to Hüber's observations, that the queen bee only lays
eggs of drones when declining nutrition or exhaustion has set in, and that
when the workers in bees and wasps lay eggs, these produce drones.
   These statements are not entirely justified. I cannot speak positively of
Fabre's observations, though I suspect something back of the larval food-
supply which has fixed the sex and determined the treatment of the larva.
But the queen bee produces drones at any age by the egg passing into the
drone cell and not being impregnated in passing the spermotheca. She produces drones only when she is superannuated, because the spermatozoa have
become exhausted. In wasps it is just the contrary, the unimpregnated egg
producing ordinarily, not a drone or a male, but a female. I have already
called attention to the ease with which erroneous conclusions are drawn in
this matter of regulating sex by food of larvæ, *ex ovo* (Am. Naturalist, Vol.
VII, pp. 513–531, September, 1873) and the evidence would seem to
how that the influence is confined to arrestation or modification of the sex
without changing it. The subject is, however, most intricate, and further
experimental facts are needed. Spencer's conclusion is, nevertheless, generally true, namely : "*that one set of differences in structure and instincts is
determined by nutrition before the egg is laid, and a further set of differences in
structure and instincts is determined by nutrition after the egg is laid.*"

*See notes on *Tiphia inornata*, Sixth Report on the Insects of Missouri, p.
123, and upon Blister-beetles, First Report U. S. Entomological Commission,
pp. 295–302,

ished one-half or more, or fully doubled, from the normal, by limiting or increasing the supply of food, as I have proved with Pelopaeus.

But when we come to the facts in the economy of the Termites, this explanation does not hold good to the same degree. Here we find still greater diversity in form than even among ants, under circumstances where control of these forms by. the colony itself must be much less, but nevertheless does occur. The young Termite is to a limited extent, and during early life only, provided with food by members of the colony, and from birth is essentially a free moving agent, less dependent on the adults. We have much yet to learn as to the actual facts, which would seem also to vary in different species. Thus in Eutermes Mr. Hubbard believes, but I think wrongfully, that the young feed on nodules, specially prepared, of comminuted and doubtless partly digested material, while Fritz Müller believes that they feed on a fungus mycelium which develops on such prepared substance. The truth with most species seems to be that they are fed on a semi-liquid fluid disgorged from the mouth, whether of the workers or the undeveloped queens; while in some cases they are fed from a secretion from the anus. (See Note 6.) In these respects and in the early helplessness of the larvæ, they closely approximate the social Hymenoptera.

Similar variations to those found in social insects, whether sexual or seasonal, are extremely common among insects which are not social, as is well exemplified by the long category of phytophagic variation, secondary sexual characters, and of dimorphism and heteromorphism among insects. These variations in non-social insects are often equally as marked and as curious, structurally, as they are among social species. They are also, except, perhaps, the secondary sexual characters and the variations which take on the form of mimicry, equally difficult to explain on any view of natural selection that is all-sufficient. On the whole, then, it may safely be said that the production of neuter insects is determined in each generation by the colony itself, in the manner in which the larvæ are fed and reared. In so far as this is true, it is outside the domain of natural selection, and speaks eloquently in favor of the various other causes of variation and modification which have been insisted upon by many of our leading American biologists, and which I have repeatedly urged in

my own writings.* The tendency to such production was doubtless developed in the ancestors of the present species, and we may
even trace the steps by studying the gradations in existing species.
The facts connected with the social insects which I have considered, present the strongest argument in favor of the heredity
of acquired characters and tendencies. Competition has been
between colonies rather than individuals, and those colonies
which have acquired, through heredity, the habit of producing,
through one or more fertile females, the different forms which
have proved useful in the social economy, have, in the course of
time, survived others in which such tendency was less pronounced. Yet various steps in the process are yet manifest
in the different species, and under these circumstances it seems
to me foolish to insist that the fixed habit in one species
has, *per se*, any especial advantage over the less fixed habit
in others which still maintain themselves. I need hardly
say to the members of this Society who are familiar with my
views as to the causes of variation, that it does not follow in
my mind that the different forms of Termites, for instance, that
are found in the colonies of some species, are all essential, but
that some of the forms may be advantageous, others only partially so, and still others purely fortuitous. The tendency to
vary—an inherent property in all organisms—has shown itself
among the individuals of these different colonies. These
variations have been guided by natural selection among colonies, and by what I have just referred to as social selection
among individuals, along certain lines which are most useful.
In other cases the variation has accumulated along lines of
secondary utility; while in yet others it has gone along lines
which are purely fortuitous and still most variable and unfixed—
natural selection playing little or no part in these. In species
with the less complete social organization, the existing variations
will be greatest; while the structures and functions have become
most fixed and show least tendency to vary in those species which
have become most specialized and perfect in their social economy.
It is very questionable, however, whether, in the struggle for existence, this greater specialization and fixity give the species any

*See more particularly the address before Section F, at the Cleveland
(1888) meeting of the A. A. A. S , and the paper before this Society "On
the Interrelations of Plants and Insects," Vol. VII, pp. 81–104 (May, 1892).

advantage over another which is more elastic and variable. On
the contrary there are many facts which go to show that extreme
specialization is a disadvantage and the precursor of decrease and
ultimate extinction. So that natural selection, in this light, if
limited, as its exponents have limited it, to the production of
characters absolutely essential or useful to the species, must play
a yet more restricted part in organic variation than even I have
allotted to it. Social selection, as here expounded, implies, it is
true, a degree of intelligence which has unusually been denied
these creatures; but the phenomena are some of them inexpli-
cable upon any other theory, and I have, I hope, already shown
how little reason we have for denying them such intelligence.

In a certain way the production of these specialized individuals
in a colony of insects may be likened to the production of
specialized individuals in a human community. In new countries, like our own, the specialization has not become so marked,
but in the older communities of the world, the life of the individual, and especially the early training and environment, produce
certain characteristics which permit us to stamp at once the typical sailor, soldier or butcher, the various artisans and the men
of various professions. They undergo essential modifications in
mind and body. Yet there is no question—or very little—of
selection, whether natural or artificial. The tendency to vary in
given directions becomes fixed through heredity, since the characteristics of different nationalities in comparison with each other
cannot be so well explained upon any other view. Certain types
persist, and the same laws which will explain the recurrence and persistence in a promiscuous community of, say, the
red-headed type, whether that of atavism or any other be ad-
duced, will undoubtedly apply to the persistency of types
in the social insects. That no material or mosaic theory of
heredity yet propounded is satisfactory, as accounting for the
facts, does not affect the question, and that natural selection,
as expounded by Weismann and the ultra-Darwinians, fails to
explain the phenomena, is the very best evidence that too much
is claimed for the theory.

INVERTEBRATE VS. VERTEBRATE.

I used to be fond of speculating as to the possibilities of the
articulate type as exemplified in the ant, in comparison with the

vertebrate type as exemplified in man, had the former continued
its development so as to approximate, say, the eagle in bodily
size and man in brain development. That the Arthropod type
could attain to such dimensions is evidenced in the Euryp-
terus or water scorpion which prevailed in early geologic times,
and attained a length of six feet; while a modern Japanese crab
(*Megachilus kæmpferi*) has a spread of ten or twelve feet, and is a
formidable creature.

For very much the same selfish reasons that begot most of our
earlier notions as to man's origin and place, it has been assumed
that he represents the perfection of the animal organization, the
highest expression of an all-wise Creator. Following this same
idea, our own world, it has been reasoned, is the only one peopled.
Now it has never seemed to me that there was any justification
for the assumption that existing forms of plants or animals must
of necessity have assumed the physical or mental characteristics
which belong to them, considering the myriad forms which have
preceded us and gone, or the many which are yet with us, but
fast going. Remembering, also, that the race is not always to the
swift, nor the battle to the strong, there would seem to be no
valid reason why, on some other sphere, under like, or even
under unlike conditions, life may not have taken on other
distinctive types or attained developments inconceivable to us ;
or, for that matter, why it might not have been differently mani-
fested upon our own little earth.

Place the directing enginery of the human brain in a body with
a hard, external skeleton, which should at once be a defensive
armor against exterior attack, a protection to all the vital organs,
and yet allow free play to every possible movement; with a
breathing system that is multiple, and therefore less liable to get
out of order than where it is concentrated in one place; with six
or more legs; extremities variously differentiated, so as to enable
one pair of them to perform the functions of our hands, while
other pairs possessed greater prehensile, tactile or other specialized powers; with powerful primary and with supplemental jaws;
with all the senses and sense organs we possess and others added ;
with simple and compound or telescopic eyes combined in the
same individual; with a venomous, offensive and defensive
weapon ; with a social organization in which working, fighting
and reproductive elements are well differentiated and yet under

control; with the power of aërial flight developed when wanted; with a reproductive system that permits of great prolificacy and yet avoids all the dangers of placental birth; with the power of temporarily suspending the active life functions when necessary; and, finally, with the power of such renewal of both the softer and harder tissues of the body as ecdysis involves—and you have in fancy a creature which would easily make the earth and all the fullness thereof its own.

The great industry exhibited by social insects has been a favorite topic wherewith to point a moral to the sluggard; but I venture to suggest that their economies, if they do not point other morals, are extremely suggestive to man. With all their other traits, so comparable to those characteristic of human society, they will hardly be charged with the possession or practice of any theology; yet we may look in vain, among all the nations of the earth, unless, indeed, among the similarly unblessed aborigines of Borneo and some other lands, for greater self-sacrifice or courage in defending the common weal; for greater loyalty to the sovereign head of the community, not made by divine right, but practically chosen by the commoners; for greater attention or care in the education of the helpless young, or for more harmonious or friendly action between the individuals that form the community. Without form or ceremony they have developed an altruism which with us is believed to exemplify the highest phase of civilization.

Nor am I quite sure that they have not solved the social problem in a way that, so far as the good of the community as well as the individual is concerned, has marked advantages over the many varied attempts in the same direction by mankind in different parts of the world. If a large proportion of the units of both sexes which go to make up human society could be so brought up and trained that the sexual instincts remained permanently arrested and undeveloped, while along with this arrestation in this particular there went an increasing intellectual development and energy, to be expended in profitable industry, what a large share of vice and misery in human society might be avoided, and what a large amount of increased happiness among the multitude might thus be secured, since in the end, intellectual and bodily activities, freed as far as possible from all baser passions, bring us the highest happiness that we can realize!

APPENDIX.

NOTE 1.—The principal Races of Apis mellifica.

The common form of this species, known as the Brown, the Black or the *German* bee, is the best-known. It is found throughout northern Europe, and as far south as central Austria, central Switzerland, and southern France to the Italian frontier. It also occurs in Portugal and Spain, and extends into Siberia, and, during later centuries, has been introduced into North and South America, many of the Pacific islands, and into Australia.

Its chief merits are that it has a moderate swarming propensity and is an excellent comb-builder and honey gatherer, and accommodates itself to the greatest extremes of climate. Its disadvantages, as compared with some other varieties, are a disposition to rob, to attack persons who approach the hive and to be somewhat less industrious. The general color is a dull brown, lighter on the thorax, the queens nearly black.

The *Heath* and *Brabant* bees, sub-varieties, occuring in the heath districts of northern Germany, are much given to swarming, a habit which has become fixed by the stimulative feeding in spring practised by the bee-keepers there for at least two hundred years.

The *Italian* or *Ligurian* bee, originally confined to Italy, Sicily, Sardinia, the southern Tyrol, and southern Switzerland, has now been introduced into most countries where the common black bee occurs. It is gentler in disposition, but not so good a comb-builder and, with a more tender constitution, does not thrive in extreme northern climates.

The color of the Italians is in general much brighter, and the first three segments of the abdomen are golden-yellow on their dorsal surfaces. Its qualities and its color have become fairly well fixed by artificial selection which there is every reason to believe has been practised in Italy for some two thousand years. Both Virgil and Columella evidently refer to it, the former (Georgics IV, 98) speaking of two kinds of bees, the better of which he describes as having shining bodies, variegated like drops of gold. The tendency to vary under domestication at the present time would indicate that the the race is a composite one, and Mr. Frank Benton informs me that by crossing the Egyptian, the Palestine or the Syrian with the common brown German race, workers are produced in a few generations that can scarcely be distinguished from Italians; a fact which as regards the Egyptians, was ascertained by the Berlin Acclimatization Society which, some 30 years ago, experimented with the honey bees native to Egypt, and which Mr. Benton has since confirmed by tests with the other two races (Palestine and Syrian) He finds also, that the Syrian tpye leads, when crossed with the common brown race, most commonly to the Italian type, a fact which is significant when we remember that the Phœnicians—ancient inhabitants of Syria—established colonies in southern Italy at a very early date. We can hardly realize to-day the importance that was attached to the production of honey and wax in Egypt and the surrounding countries in those days, until we remember the uses to which these articles were put in connection with the religious rites of the people, and especially the embalming of the dead, as well as the relative importance of honey in those early days in the absence of the many other sweets which we possess. In the United States the Italian race, by selection since its introduction a third of a century ago,* has undergone more rapid modification than any of the other races, though

*See a paper by the author on "What the Department of Agriculture has done for Apiculture." Proc. North American Bee Keepers' Association, 1893.

greater efforts, proportionately, have been made with these—another fact which would indicate that the Italian type is less fixed than some of the oriental races.

The *Carniolan* race is confined to Carniola, Austria, and the adjoining provinces, and is a local type developed by some centuries of peculiar treatment with little intermixture of outside blood. This race is somewhat larger than the others, exceedingly robust, the distinctive color-mark being light gray varying to steel blue, the abdominal segments being all edged with pubescence of this color and the thorax thickly set with the same. The race is characterized by great prolificacy, which can be traced to the constant stimulative feeding early in the season, and by a very mild disposition, a result which would seem to be due to the frequent manipulation of the hives, migratory bee-keeping having been practised for centuries in Carniola.

The *Cecropian, Attic,* or *Hymettus* bees of Greece, on the other hand, though similar to the Carniolan race in markings, are exceedingly irritable, as a result, doubtless, of their being very little manipulated or interferred with.

The *Tunisian* bees are found in Tripoli, Tunis, and Algeria, where they are extensively cultivated by the natives. The type is uniformly dark in color. The queens are very prolific and when preparing to swarm 200 to 300 queen-cells are often constructed, instead of only 8 to 10 as is usual with the ordinary race. The workers are small, very active, irritable and vindictive. Because of this and the fact that they do not winter well, in consequence of prolonging the brood season, their introduction has been very limited.

The *Egyptians,* or the bees found all over northeastern Africa, and which for several thousand years have been extensively cultivated in Egypt, possess very marked characteristics as regards color, form and habits, and have been regarded by many as worthy of specific rank, having been described by Latreille as *Apis fasciata.* The workers are small-bodied, slender, covered with a dense, light gray pubescence, and the abdominal segments are edged on their dorsal surfaces with a lemon-yellow color, giving with the gray pubescense a banded effect. They do not withstand our winters and are easily angered by manipulation, not being amenable to smoke like European bees. The queens are prolific and when the colonies are made queenless great numbers of workers commence depositing eggs at once.

The *Palestines* and *Syrians* possess many of the qualities and characteristics of Egyptians; yet the queens, workers and drones are readily distinguishable from those of the latter, being less yellow and larger bodied, especially the Syrians. They are marked varieties, more fixed than the Italian, and evidently forming, with other eastern Mediterranean bees, an Oriental group having allied characteristics and of which the Egyptian is the extreme type.

The *Caucasian* and *Smyrnian* races vary more than the other Oriental races. In specimens from Smyrna the light yellow coloration of the abdominal segments noted farther south is found to be replaced by a darker yellow and the light gray pubescence by a less dense and darker gray, often brownish, pubescence. Queens, workers and drones are larger bodied and variations in temper and habit may also be noted.

The *Cyprian* race, having been isolated for a long period, is, as might be expected, a very fixed one—the most thoroughly so of any race of bees yet brought to this country, and transmits its peculiar markings and characteristics through many generations of crosses with any other known type. In general it resembles the race found on the adjacent mainland, whence it was probably brought by the early Phœnicians who colonized Cyprus. Very characteristic markings of this variety are the bright yellow lunule which the postscutellum shows and the bright yellow of the ventral surface of the abdomen clear to the tip. The conditions under which this race has been established have resulted in the survival of a hardy, active race, capable of procuring a living and storing a surplus where others could barely subsist.

The literature refers almost entirely to the older countries of Europe and the East. Some modification has doubtless taken place in the tropical parts of America but the subject has not yet been sufficiently studied in those countries.

NOTE 2.—The Species of Apis with their Varieties.

(1) *Apis mellifica*, L. as indicated in Note 1, is found in all the countries of Europe, and extends over the whole of Asia Minor into the Syrian Desert and south into Arabia. It occupies all the islands of the Mediterranean and has spread through all the northern countries of Africa southward into the Desert of Sahara. South Africa has one or two varieties belonging to the species, while the representatives of the genus found in Senegal and the Congo country doubtless belong to this species, as do those of Madagascar. It has been permanently introduced into Australia, Tasmania, New Zealand and many of the islands of the Pacific ocean. Whether the honey bees reported from northern India belong to this species or not, has not been definitely ascertained. It is also more than probable that the honey bee of China, described under the name of *Apis sinensis*, is but a variety of this species. In North and South America it is evidently introduced, and has spread into some of the adjacent islands. There is a difference of opinion as to whether the honey bee native to Egypt, which Latreille describes as *Apis fasciata*, should have specific rank or be regarded as a variety of *mellifica*. While Frederick Smith, who was one of our best authorities, was inclined to attribute to it specific value, the fact that it interbreeds with *mellifica*, producing fertile offspring, would rather confirm the opposite view. Respecting the honey bees of Tasmania, Senegal, the Congo and Madagascar, our information is insufficient to permit us to say whether they are specifically distinct or not, and the same may be said of the Hazara, Bhootan, and Bashar bees of northern India and other more or less distinct types found in Japan.

(2) *Apis indica* Fabr. The extent of territory occupied by this small East Indian bee is not definitely known, although it has been definitely reported from northern and southern India, Ceylon, Farther India and Java. *Apis nigrocincta*; *A. socialis*, Latr.; *A. delesserti* Guer.; *A. perrottetii* Guér. and *A. peronii* Latr. are probably only varieties of *A. indica*.

(3) *Apis florea* Fabr. This, the smallest bee of India, is found generally in southern India and Ceylon, and there are indications, that it is common to other portions of the East Indies. *Apis lobata* described by F Smith in his first catalogue, is dropped from the second edition.

(4) *Apis dorsata* Fabr.
> =*nigripennis* Latr.
> =*bicolor* Klug.
> =*testacea*.

It is somewhat questionable whether the names here given as synonymous are such, or names of true varieties of *dorsata*. *A. dorsata*, known as the Giant East Indian Bee, is found in British India, Ceylon, Farther India and the Dutch East Indies.

(5) *Apis zonata* Guérin. Found in the Philippine Islands and Celebes. Mr. F. Smith enumerated this as worthy of specific rank, when he revised his catalogue in 1876. He referred to its greater size and difference in form of the metatarsus compared with that of *A. dorsata*. But Gerstaecker asserted in 1865 that this difference in structure of the metatarsus does not exist—is "purely imaginary"

Mr. Frank Benton, to whom I am under obligations for valuable information on this subject, has kindly prepared for me the following table as indicating his own ideas of the grouping of the species of Apis, and the known varieties of these.

The Species of Apis with their Varieties.

Apis mellifica* Linn.

Race.—Common Brown, Black, or German.—Hab.: Central, northern and northwestern Europe; introduced into N. and S. America, Australia, New Zealand and Pacific Islands.
    Sub-var.—Heath.—Hab.: Heath districts of North Germany.
    Sub-var.—Brabant or Small Holland.—Hab.: Brabant (Holland and Belgium).
Race.—Carniolan.—Hab.: Carniola, Carinthia (Aus.). A distinct var.
    Sub-var.—Hungarian.—Hab.: Northwestern Hungary.
Race.—Dalmatian.—Hab.: Dalmatia (Austria).
Race.—Herzegovinian.—Hab.: Herzegovina (Austria).
Race.—Cecropian, Attic or Hymettus.—Hab.: Greece and the adjacent islands.
Var.—*ligustica* Spin., Ligurian or Italian.—Hab.: Italy and adjacent islands. S. Switzerland, and S. Tyrol; introduced into other parts of Europe, N. and S. America, Australia and New Zealand.
Var.—†*rufescens*—Hab.:—Tasmania (acc'd to M. Girard).
Var.—†*nigritarum* St. Farg.—Hab.: Congo (Africa).
Var.—†*adansoni Latr.—Hab.: Senegal (Africa).
Var.—*scutellata* St. Farg.—Hab.: South Africa.
Var.—*caffra* St. Farg.—Hab.: South Africa.
Race.—Tunisian.—Hab.: Tunis, Algeria.
    Sub-var.—Minorcan.—Hab.: Balearic Islands (Spain).
Var.—†*unicolor Latr.—Hab.: Madagascar ; intr. into islands of Bourbon and Mauritius.
Race.—Smyrnian.—Hab.: Asia Minor.
Race.—Caucasian.—Hab.: Caucasus.
Race.—Cyprian.—Hab.: Island of Cyprus. A very distinct var.
Race.—Syrian.—Hab.: Syria, northward from Mr. Carmel.
Race.—Palestine.—Hab.: Palestine.
Var.—*fasciata* Latr.—Hab.: Egypt.

Apis sp.

Race.—Hazara.—Hab.: Hazara District, Punjab (India).
Var.—**sinensis.*—Chinese bee.—Hab.: China.
    =*cerana* Fabr.
Race.—Bushar.—Hab.: Bushar District, Punjab (India).
Race.—Japanese { 1. "Grayish yellow bee." 2. "Bee with yellow spots." } Hab.: Prov. Sinano. 3. "Small brown bee."—Hab.: Hikigoie (Satsuma)
Race.—Boohtan.—Hab.: Boohtan (India).
It is very probable that further investigation of this group will bring four of its varieties under *A. mellifica*, and the last one under *A. indica*.

Apis

*indica Fabr., Small East Indian bee.—Hab.: British and Dutch East Indies.
    =*socialis* Latr.—Hab.: Bengal.
    =*delesserti* Guér.—Hab.: Pondicherry.
    =*perrotteti* Guér.—Hab.: India.
    =*peronii* Lat.—Hab.; India.
Var. (?)—**nigrocincta.*—Hab.: "Celebes, Borneo, etc." (acc'd to F. Smith.

---

*Regarded by Frederick Smith as good species.
†Not positively known that they will interbreed with **Apis mellifica**. All others named under *A. mellifica* will do so.

Apis { **florea* Fabr.—Hab.: India, Ceylon, Borneo.
  =lobata Smith.—Hab.: India.

Apis { **dorsata* Fabr.—Hab.: British India, Ceylon, Farther India,
      Dutch East Indies.
    =bicolor Klug.
  Var.—*nigripennis* Latr.—Hab.: Bengal.
  Var.—*testacea* Smith.—Hab.: Timor.

Apis { **zonata* Guér.—Hab.: Philippine Islands, Celebes.
  This may prove but a variety of *A. dorsata.*

NOTE 3.—Polliniferous Organs in Bees.

The modification of structure and hairy vestiture (see Fig. 2) to facilitate the collection and transportation of pollen is, perhaps, exhibited in its most perfect development in the Hive Bee. That these peculiarities have been evolved from those possessed by less specialized species of social bees, represented by existing Meliponae and Bombi, and still more remotely from those of solitary bees, will not be questioned by those who study the steps in the process as exemplified in modern species.

The pollen of flowers is variously collected by different bees, and different parts of the body are specially developed for this purpose. But in the Hive Bee the specialized polliniferous apparatus is limited to the posterior legs, and in these to the tibia and the basal joint of the tarsus, so that the development of these parts only need be traced.

In the case of the tibia the first thing to be noted is the entire absence of the tibial spurs, which are present in all Hymenoptera except the genus Apis, and its near allies Melipona and Trigona. The tibia and first tarsal joint are greatly broadened and more or less concave exteriorly, and the latter is extraordinarily enlarged, so that it is nearly equal in size to the tibia. The outer surface of this modified tarsal joint is not remarkable and has no specific function, but the inner surface is divided into transverse rows of stiff spines or combs, reddish in color, the rows slightly overlapping and elevated at a slight angle from the surface of the joint. The function of this series of combs is to collect the pollen grains which become entangled in the feathery hairs of the thorax of the insect, and an examination will almost invariably discover more or less of the grains of pollen in these combs. During the collecting of honey and pollen, the bee is constantly passing the face of this tarsal joint over its abdomen, removing the pollen grains from time to time, and emptying the load of pollen into the pollen-basket proper or corbiculum, on the outer face of the tibia. This, as noted, is concave, with a smooth, almost hairless exterior surface, provided at the sides with several rows of long curved hairs, which arch over either side, forming a veritable basket in which the pollen may be securely packed. As soon as the collecting combs of the tarsus are filled, the bee draws them across the strong, curved hairs of the corbicula, the right tarsus emptying into the left corbiculum and *vice versa*, until both are filled. These baskets or masses of pollen are emptied by means of the single strong tibial spine on each of the middle pair of legs, the spine being thrust beneath the load of pollen and used as a pry to loosen and remove it.

A very remarkable peculiarity of the posterior legs, but having no connection with the polliniferous apparatus, is seen at the union of tibia and first tarsal joint. These are articulated at the extreme anterior angles in such a manner that the broadened apex of one and the base of the other, work together as a sort of nippers or pincers. The tibia is armed on the inner margin with a strong, uniform row of short spines extending two-thirds of the way across. This apparatus is employed by the bees in removing the wax scales from the abdomen.

Examination of these parts in other species of Apis fails to indicate any particular modification or deviation in structure from *mellifica*. In *Apis indica* no differences whatever can be discovered; in *A. dorsata* the leg is somewhat more hairy and a few hairs occur on the outer surface of the tibia. In *A. florea* the smallest species known, the spines on the apex of the tibia are somewhat shorter and stouter and the hairs forming the corbiculum are somewhat less regular in length and arrangement.

This statement of the structure of these parts in the species of Apis will enable us to compare intelligently the similar parts in those genera most nearly allied to them, tracing the variation through these to the more widely divergent forms. The genera Melipona and Trigona include bees which are closest to Apis in general structure and habits, and agree also in the absence of the tibial spines of the posterior legs. We find, as might be inferred, a very close correspondence in the polliniferous apparatus, which, in all essential details, is practically the same as in Apis. The pollen-collecting combs on the inner surface of the first tarsal joint are absent, or rather their place is supplied by a uniform clothing of short stiff spines which are not arranged transversely in rows, as in Apis, but serve the same purpose. This joint also differs in shape from that in Apis, by being suddenly narrowed or excavated toward the base so that the nippers noted in the former genus for the removal of the wax are practically wanting, although the row of stiff spines at the apex of the tibia is still present, but somewhat reduced. A very peculiar tuft of strong, curved spines occurs, in the two genera mentioned, at the anterior outer angle of the tibia. This has no counterpart in any other bees and its function is problematical.

In the case of Bombus, the lowest of the social bees, there is at once a greater divergence from Apis and at the same time a resemblence to it in certain features of the hind legs and polliniferous apparatus. The tibial spines are very strongly and prominently developed, allying this genus to the solitary bees and other Hymenoptera, but the general structure of the tibia and first tarsal joint is practically identical with that of Apis, and the tarsal joint in this particular does not present the divergence which was noted in the case of the genera Melipona and Trigona, but has the broadly truncated basal margin forming the lower blade of the nippers, even more strongly developed than in Apis. The pollen-collecting spines on the inner face of the tarsal joint are uniformly distributed over the surface, practically as in the two genera last mentioned (Melipona and Trigona). The bordering hairs of the corbiculum are somewhat stronger and more abundant, but in all essential details the structure is identical with the same in Apis.

The solitary bees of the genus Anthophora, which is somewhat nearer Apis than any other, present distinct traces of the specialized polliniferous apparatus of this last. The enlargement of the tibia and of the first tarsal joint is quite marked, and the corbiculum is imperfectly indicated by the longer growth of hairs on the edge of the tibia, the face of the latter being also covered with shorter hairs. The brush or pollen comb on the inner surface of the tarsal joint is practically the same as in Bombus. The small row of spines at the apex of the tibia are entirely wanting, and the nippers at the junction of the tibia and metatarsus are not particularly noticeable; in fact this structure is not seen in any except the social bees which alone produce and use wax in their economy. The genus Melissodes presents a distinctly wider divergence from Apis, in that the hairy vestiture on the outer surface of the tibia and metatarsus is equally long and dense over the entire surface, showing little if any approach to the corbiculum, which, as we have seen in Anthophora, begins with the shortening of the hairs on the outer face of the tibia. In other particulars the bees of this genus are similar to Anthophora, and in both genera the pollen collected is carried interspersed among the hairs of the tibia and tarsus, being doubtless emptied or combed into them from the brush of the inner surface of the first tarsal joint, and probably removed again by the same brush in storing it in their larval cells.

Going still lower in the scale of bees, we find in Perdita a yet wider divergence from Apis in the absence of any particular dilation of the tibia and metatarsus, the posterior legs being similar to the anterior members, simple in structure, and armed with long, scattered, feathered hairs, which are generally distributed over all their surface and which entangle more or less of the pollen grains. The brush of the inner surface of the metatarsus is still present, and in fact occurs in all Apidæ and Andrenidæ. The genus Nomada is still less specialized, in that the legs are simple, not dilated and also practically hairless; or rather the hairs which are short and simple and have no pollen-collecting capacity. In this genus the brush of the metatarsus can hardly have any other use than to keep the body of the insect clean, as these bees are pseudo-parasitic or inquilinous and do not collect or store pollen. It is a mere modification of the normal or original structure and doubtless a degeneration due to the semi-parasitic habit.

From the above review of the modification of the posterior legs as pol-liniferous organs in various genera of the family Apidæ, it will be seen that there are first developed on the leg, hairs which are feathery and which will entangle the grains of pollen. The next step in the development is an increase in the abundance of this hairy vestiture, and a further advance occurs in the widening of the tibia and first tarsal joint, to give a greater surface for the pollen-collecting, plumose hairs. This reaches its maxium in the genus Melissodes in which the external hairs of both the tibia and the metatarsus are very long and dense and the feathering very decided. The next step toward the condition found in Apis is exhibited in Anthophora, and consists in the partial disappearance and shortening of the hairs on the outer face of the tibia and metatarsus, by which means an imperfect corbiculum is formed, foreshadowing the more complex structure of the social bees, in which it becomes quite well developed in Bombus and perfectly so in Trigona, Melipona, and Apis. In Anthophora a further modification is noted in that the hairs of the legs are practically simple and unfeathered as in the higher social bees.

In the other family of bees, the Andrenidæ, we have a similar condition of things, the variation in the pollen-collecting character of the posterior legs ranging from Agapostemon to Prosopis, and showing the same gradations noted in the Apidæ from Melissodes to Nomada.

The reader interested in studying how the mouth-parts and the legs have been modified in the bees by their honey and pollen gathering habits, cannot do better than consult Hermann Müllers' works* on the subject. There is almost an unbroken chain of these characters, from the highly developed bees to such as are hardly distinguishable from the fossorial wasps.

NOTE 4.—Wax-producing organs.

In all the wax-producing bees the specialized discs (see Fig. 3) on which the wax is deposited when secreted by the true glands beneath, occur on the basal half of the second to the fifth ventral segments of the abdomen, the overlapping half of each segment covering and protecting the disc of the succeeding segment. With the Hive Bee these discs are compound and two in number on each segment. They are broad, ovate, pale yellow in color, smooth, delicate and transparent, and are surrounded by a narrow thickening of the chitine of the sclerite and separated by an unmodified medio-ventral septum. This specialized structure occurs only in the workers. The queen, however, has a sub-obsolete, undivided area on the same five abdominal segments, and which in structure bears a striking resemblance to the similar area in the workers of the lower forms of social bees. The wax discs of Melipona and Trigona are practically identical, and are narrow, extending entirely across the base of the segment, not being broken, as in Apis, with a dividing septum, and also extending laterally

---

*The Fertilisation of Flowers, by Prof. Hermann Müller. Translated and edited by D'Arcy W. Thompson, B. A., London, 1883.

nearly to the apex of the sclerite as in the case of the fertile female in Apis. In Bombus the structure is almost identically the same as in Melipona.

Note 5.—Ant Economy.

Considering the large number of species of ants, a book would be required to treat of them in detail, and volumes have been written. In this note I shall only treat of a few of the better known, to supplement the mere summary in the body of the address. The most interesting of our North American species which I have had an opportunity of studying are the mound-building species of the East, the leaf-cutting species of Florida and Texas, and the honey ants of Colorado. With the aid of Mr. Th. Pergande, who has been assiduous in his studies of the family, and is perhaps our best-informed myrmecologist, I have brought together a number of notes on the habits of our North American species of Carpenter Ants and others; but they are excluded as the least important in connection with the text, and with a view of duly limiting the pages.

MOUND-BUILDING ANTS.—In this category may be classed by far the larger number of our better-known ants. The term is, however, particularly applicable to the species of the genus Formica. These ants are very much more active and industrious and typical of the family, than are the carpenter ants. Our own species inhabit, by preference, pine woods. They are pugnacious and valiant, and whenever their mound is disturbed, however slightly, will speedily cover the whole surface in one surging mass, spreading over the mound and attacking in their fury any living creature within reach. They are in fact so fierce and fearless that even man does well to avoid their mounds; for the bite is quite severe, and when multiplied indefinitely is unbearable.

The Fallow ant (*Formica exsectoides* Forel), one of our best known species and a close ally of *F. exsecta* of Europe, builds large mounds of earth, more or less mixed with other materials, especially small sticks and dried leaves of pine. These will measure all the way from two to eight feet in diameter at the base, and may be from one to three feet high. They are more or less regular and conical, full of galleries, with larger or smaller chambers which communicate with a general system of subterranean cells or cavities, which are used as store-rooms, nurseries for the young, parlors for the queens, and other purposes. The purpose of the superstructure in most mound-building ants appears to be for aëration, for the more rapid development of the larvæ, and, apparently, to facilitate social intercourse between the individuals when not engaged in actual work. Except for the extraneous matter which gives it firmness, all the material of the mound is brought up from beneath the surface, and the inhabitants are incessantly at work, night and day, in constructing, altering and repairing. Very large colonies are often connected by secondary hills. I once had a good opportunity of studying these mounds around Ithaca, N. Y., and Dr. H. C. McCook has published a most interesting and detailed account of his observations upon this ant in the Trans. American Entomological Society for 1877, Vol. VI, page 253, and also in *The American Naturalist* for July, 1878, Vol. XII, pp. 431–445. It is particularly common in the Alleganies. There are three forms of workers, viz, major, minor and dwarf. His interesting observations will well repay reading.

It is in these mound-building ants that we find the true economy of the division of labor. While large numbers are ceaselessly building and mining, so as to keep the formicary in good condition, repairing or increasing its size, so as to accommodate the growing numbers, others are busily engaged in scouring the surrounding country for food, both for themselves, for the multitude of those who stay at home, and for the young. In these expeditions they never hesitate to attack any other insect that may be in their way, no matter how much larger than themselves, and what they lack in power individually they make up in numbers. Still others again are running over the trees and shrubs and other plants, searching for plant-lice, from which they gather the sweet rejectamenta, gorging themselves frequently to such an extent that they return home with difficulty. This honey is used chiefly for feeding the larvæ.

HONEY ANTS.—There is really but one Honey Ant, strictly speaking, viz, *Myrmecocystus melliger* Llave (*M. mexicanus* Westm.), in North America, and this ranges from Mexico to Colorado. Other species occur in other parts of the world, with somewhat similar habits, and one is especially mentioned by Lubbock from Australia (*Camponotus inflatus* Lubb.) which has undergone precisely the same modifications, though belonging to a distinct genus, a most interesting fact, since it shows that the modification has arisen independently. The honey collected and stored by these ants has little value commercially, first, because of its rather poor quality; secondly, because of its small quantity—barely more than half a pint to each colony—obtainable; and, thirdly, because of the difficulty of colonizing or in any way commercially manipulating the ants. The insect must be crushed to obtain the honey. Yet it is sought for by the Mexican Indians, and used to a considerable extent. The formicaries are little truncated cones from two to three inches high, and usually less than a foot in diameter. They have a tubular channel, a few inches in diameter, leading from the central opening to the interior, to a depth of six inches or more below the general surface. Here are often found one or more dome-like vaults or honey-chambers, about an inch deep by about three inches in width. Hanging from the roughened roof of these chambers may, at any time, be found numbers of the honey-bearers, with immensely swollen abdomens and looking, when congregated, like a series of small grapes or large currants, with the same translucency which these possess. These individuals have little capacity for movement, and indeed move but little. They are but living receptacles of the sweets which are gathered by the real workers, and the food-supply of the rest of the colony is only drawn from these stationary honey reserves, or animated honey pots, as Lubbock calls them, when necessity requires. The modifications are confined to the abdominal portion of the digestive organs. The honey is gathered from a little Cynipid oak-gall which I have described as *Cynips quercus-mellaria* and which abounds on a small scrubby oak (*Quercus undulata*) frequent in those regions. The ants always work at night, making their way in long strings to the nearest gall-bearing tree, the branches of which they carefully search for the young and succulent galls which secrete a small globule of a clear saccharine liquid. The gathered liquid is then, upon the return to the formicary, emptied into the mouths of those individuals which serve as honey stores.

LEAF-CUTTING ANTS.—These are represented almost solely by the genus Atta, which abounds in tropical and sub-tropical countries, where the species are dreaded by planters because of their great destructiveness to cultivated plants and trees. These ants have been denominated agricultural ants, and recent observations have confirmed the explanation originally urged by Belt, that the leaves are cut into pieces and gathered into small heaps, as a nidus for the cultivation of a fungus (Rozites) the mycelium form of some mushroom, so that they may be said to have anticipated man in this kind of culture. The only two species belonging to the genus so far observed in this country, are *Atta fervens* Say, and *Atta tardigrada* Buckley. The former is our commonest species, occurring in Texas. Its formicaries are often twenty feet in diameter and several feet high, with numerous smaller moundlets scattered over the surface. They have a crater-like depression in the top, with a central opening running down into the formicary, sometimes to a very great depth. Each formicary contains immense numbers of individuals, and during the day appears to be empty and deserted. After dark, however, the entrances are opened, first by smaller workers who remove the particles of sand and earth, then by individuals of larger form who aid in removing the refuse. When the way has been sufficiently cleared, the inmates pour forth, both workers and soldiers, and march to some plant or other near by. They are generally seen in double column, one column ascending the plant and cutting off the leaves, and the other returning loaded to the nest. Great intelligence is shown by this ant in its foraging expeditions. The cut leaves, either whole or in circular pieces, are usually thrown on the ground by those who ascend the tree, while others below receive and bear the fodder home. Each piece of leaf is grasped by the jaws, and, with a quick motion of the head, thrown back over the head and thorax in such manner that it lodges edgewise in a deep furrow and between two spines which characterize the head, so as to cover the insect more or less and offer little or no obstacle to its progress. Very long underground tunnels are sometimes excavated from the main formicary to some shrub or tree so as to facilitate access thereto. The stories told by southern planters of the ravages of this insect seem almost incredible, but I have myself witnessed the utter denudation of a large tree in a single night, in which case all the forces of the formicary seemed to be concentrated on a single object.

Atta tardigrada is found east of the Mississippi River, occurring throughout the gulf States from Florida to Texas. In Florida what is evidently this species builds rather large cells from two to four inches in diameter in fine white sand, the walls very firm and smooth. In some instances the walls are said to be lined with a kind of curtain composed of particles of different colored sands brought up from a lower stratum and interwoven with fine white threads, by which is doubtless meant shreds of the refuse vegetation collected—a kind of spongy mass, manufactured from the vegetation and somewhat resembling the comb made by certain bees. This spongy mass contains small irregular pockets, apparently designed for the reception of the young, and in this we have the nearest tendency in ants to the building of cells which is so common in some of the other social Hymenoptera. This species prefers the fine needle-like leaves of tender pine seedlings, and a row, marching in single file, each carrying a piece of one of these needles, suggests a file of soldiers armed with rifles.

Atta mexicana Sm. abounds in the temperate regions of Mexico, its formicaries being twenty or more feet in diameter, and a funnel is said to extend through its center to facilitate drainage, which would seem to be necessary in a country subject to very heavy rains. The damage done by this species, especially to coffee plantations, is said to be very great.

Atta cephalotes L. is dreaded in Brazil because of its destructiveness to vegetation and of its tendency to enter houses and carry off the mandioca meal. Its formicaries often reach a diameter of more than 100 feet.

NEST-BUILDING ANTS.—Though we have in the United States no species which constructs nests similar to those of wasps, yet such are known to occur in other parts of the world, especially in tropical and sub-tropical countries. The genera Polyrhacis, Dolichoderus and Cremastogaster imitate wasps in the construction of their nests.

Some of the Brazilian species of Cremastogaster construct more or less globular, black nests, about the size of a human head, fastened between the branches of trees, large numbers of which may often be noticed among the mangrove bushes bordering the shores of the ocean, and frequently so low down as to be but a few inches above high tide. Similar nests are common in the West Indies, and look very much like young nests of Eutermes.

The nest of *Cremastogaster arboreus* Sm., found at Port Natal, Africa, is very large, measuring about fifteen inches in length, by nine inches in diameter. It is always built around a branch, resembles in texture and appearance the nest of our common paper wasp, *Vespa maculata*, and contains thousands of the insects. (See Smith, Cat., Hym. **Ins.** Brit. Mus. Pt. VI, pl. XIV.

We see the beginnings of the nest-building habit in some of our North American species, especially in *Cremastogaster lineolata* Say, which builds coverings over colonies of Aphides, the coverings composed of minute particles of vegetable and earthy matter firmly glued together ; or else makes a more or less conspicuous loose nest by massing together the exuviæ of the Aphides and portions of dead leaves, generally around some twig or branch. (See *Practical Entomologist*, Vol. II, No. 3, Dec. 1866, p. 41.) In this case the object is doubtless to prevent the robbing of the coveted sweets by other nectar loving species; while the more elaborate nests of the tropics are for self protection and social economy, the nearest approach to these in N. A. being made by a Florida ant (*Cremastogaster lævinscula* Mayr) which makes large brown chambered nests in long grass, recalling somewhat in color and character those of Eutermes.

Note 6.—Termite Economy.

TRUE ROYAL PAIRS.—There are many recondite phenomena connected with the life-history of the Termites that yet remain unexplained. But all the species annually produce large numbers of male and female adults, i. e., winged individuals which are capable, normally, of reproducing. These are recognizable after the first moult by the larger thoracic segments, which bear the first indication of wing-pads. During flight or swarming, and the subsequent walks on the ground, no real union of the sexes has so far been observed. In fact the reproductive organs are at this period not fully developed, and it is not until a pair have succeeded in establishing themselves amid a certain number of workers that the sexual organs become functional. The wings are thrown off and at this stage these individuals are known as true royal pairs, the wing stumps showing in contradistinction to the wing-pads of the larva and pupa, while their darker color otherwise distinguishes them. They are long-lived, coition taking place repeatedly. The male increases but little in size, but the abdomen of the female increases enormously with increasing fecundity.

SUPPLEMENTARY KINGS AND QUEENS.—The absence of a true royal pair by no means impairs the vitality and prosperity of a Termite colony; for a certain number of individuals are met with which, in the absence of the true queen may become sexually mature, the female laying fertile eggs, from which, in due course of time, all the forms composing the colony are developed. The true nature of these secondary or supplemantary males and females was first fully recognized by Fritz Müller, and their development is explained as follows :

At first indistinguishable from the larvæ of individuals which produce winged specimens, they are, in the nymph or pupa state, thicker and clumsier. The internal sexual organs are more strongly developed, and they have short wing-pads placed sideways instead of long and broad wing-pads as in the nymphs which produce the true kings and queens. In short, they undergo one moult less, and, as a consequence, do not acquire wings or swarm. They acquire sexual maturity later in the season than the winged individuals, from which they are always distinguished in maturity by the possession of wing-pads instead of the wing stumps. They are also lighter in color, the males having smaller eyes, and the females a broader thorax, whereas in the true royal individuals there is no difference in this respect. They are not as long-lived, either, as the royal pair, the males dying within a few months and the females probably not surviving more than a year.

It will be seen from the above stated facts that if through the death of a queen, or in the absence of a queen, a colony has not been able to secure another royal pair from the swarming individuals "nymph-like males and females, safely kept in the nest" step in as substitutes and save the colony from becoming extinct. Furthermore it has been observed that if, in very small and fragmentary colonies, the supplementary males and females should be absent, the colony may yet be perpetuated by the substitution of larva-like males and females, which have been called complementary kings and queens.

A remarkable observation made by Fritz Müller deserves mention here. He found in a Eutermes colony, in the passages of what appeared at first to be a true royal cell, not less than 31 supplementary females and among them a single true king. i. e., with distinct wing-stumps. "Instead of a royal palace" he says, "in which the king lived in chaste matrimony with his equal consort, I had a harem before my eyes in which a sultan satisfied himself with numerous coquettes." This observation would seem to indicate that, in the economy of the Termite colony, a true king and queen may not only be replaced by supplementary kings and queens, but that this substitution may take place for both sexes at the same time, or for each sex separately.

I would observe, in this connection, that during the swarming season many species of true ants forcibly detain some of the winged males and females and prevent their leaving the formicary by biting off their wings, and that the pairs thus forcibly detained supply the colony with eggs. A similar condition may prevail among the Termites, and if so, would throw light on some of the facts which have been observed.

INFLUENCE OF FOOD AND TREATMENT.—The effect of food and treatment has less, perhaps, to do with the differentiation of individuals among termites than among the bees, wasps or ants.

All Termite larvæ are supposed to partake of the same kind of food, as to the nature of which there is conflict of opinion, due doubtless to the varying habits in the different species. From my own observations on Termes and Eutermes, I am inclined to believe that, as in the Social Hymenoptera, the food and treatment of the young larva, during the first stage more particularly, have much to do in determining the development or suppression of the sexual organs, and, as a consequence, in determining the character of the full grown individual. The eggs are, first of all, brought together in special parts of the termitary, and it is quite probable that the workers exercise some judgment and discrimination in the grouping, as has been proved to be the case with Hymenoptera, with a view to future larval treatment. Judging from the delicacy of their mouth-parts and of the general integument, the young are at first more or less dependent upon either the forethought or the direct action of the adults, and I cannot resist the conclusion that the infancy of the termites is dependent, as it is in the Social Hymenoptera, if not to the same extent; for they have soon perished where I have hatched them away from adults, and have developed where the adults had access to them. But further exact observations, which, in the nature of the case, it is difficult to make, are needed before definite conclusions can be drawn. Fritz Müller believes that the young feed on a fungus which develops on the walls of the cells, a peculiar white fungus being not uncommon in such situations, though I have more often found nothing of the sort where the young were abundant.

Mr. Hubbard found many small hard bodies among the eggs of *Eutermes rippertii* which were recognized as the sclerotium of a fungus by Prof. F. G. Farlow, and other observers have referred to the presence of fungi in Termite nests. Mr. Hubbard also records the feeding of the young upon hard and tough rounded masses found in the nests of the above-named species. They could not do so, however, without the assistance of the nasuti or workers to soften these nodules, for their mouthparts are too feeble, while the nodules are of very irregular occurrence and in some nests not present at all. Where the young are crowding, the material of the nest is moister than elsewhere and their chief food must be a liquid regurgitated from the mouth, by the workers or by the partly developed sexed individuals, just as in the social Hymenoptera, and either taken directly or from the moistened substance of the cavities. Indeed, though Mr. P. H. Dudley in some interesting observations on Eutermes on the Isthmus of Panama (Journal N. Y. Micros. Soc. V. p. 62, April, 1889) describes the nasuti as being able to fire an "offensive glutinous shot, which puts an antagonist twice his size *hors de combat*," I have never been able to confirm this statement. The nasuti have seemed to me defenseless and I suspect that the liquid so readily secreted from the tip of the nose is chiefly designed for nourishment. That comminuted, decayed wood, as well as the fæces are also used for food has been shown by Grassi and others, while the tendency to feed freely upon one another is matter of common record, and indeed all the dead and dying are devoured.*

In Calotermes the excrement consists of dry and hard sub-ovoid particles which accumulate in the burrows, so that the faeces are not used here whether as food or to line the burrows. Consequently the young must depend entirely on liquid from the mouths of the females. The food is, however, from what has gone before, sufficiently varied in those species which exhibit the greatest number of colony forms, to justify the belief, here set forth, that it has much to do in the development of those forms.

It is, however, definitely known that differentiation of the sexes takes place at an early period, and can be recognized by anatomical and external characters in the larva, immediately after the first moult. Freshly hatched larvæ appear to be sexually undifferentiated, although it is probable, as suggested by Newman in 1853 and Hagen in 1855, that this is simply because the differences are too minute to be observed. Sex is doubtless determined in the egg, but the different forms of either sex are, in all probability, due to food and treatment in the first larval stage, and to an innate tendency confirmed by heredity. The mode of treatment of the mother, in insects generally, may influence the sex of the offspring; but there is no evidence to show that the sex can be altered when the egg has once passed. Fecundity varies in individuals of any community, and a certain number are always sterile. In the social insects this condition is simply controlled to the advantage of the species, and the tendency, associated with various other modifications, has been emphasized by heredity. Prof. B. Grassi (Bull. Mensuel Acad. Gioenia, 1889; Entom. Nachrichten, 1889) has offered a rather curious explanation of the origin of the sex in Termites. He finds in the cœcum of the young larvæ, as well as in the fully developed workers and soldiers, an abundance of protozoon parasites. With each moult these parasites disappear, but immediately commence to reappear, and the cœcum is inflated in a sac which presses on the sexual organs so that the development of the latter is prevented, the protozoons not appearing, after the first moult, in those individuals which are to become truly sexual, or at least in only the smallest quantities. He bases this view upon the examination of many hundreds of individuals, but the probabilities are that the presence of the protozoons has no essential part in the result, as he offers no explanation as to why they are absent or less numerous in the one case than in the other.

COMPOSITION OF THE TERMES COLONY.—Remembering that in Termes the adolescent stages actively participate in the work and composition of the colony, and accepting the nomenclature most recently used by the latest and best observers, the forms already indicated in the diagram on p. 33 may be enumerated as occurring in the species of the genus Termes, as exemplified by the commoner European and American species:

Prof. Grassi has enumerated some three additional forms, but this confusing complexity of forms really occurs only among those which are reproductive and they never all occur at one and the same time, while some of them only occur under certain peculiar conditions.

The youngest larvæ, i. e., the indistinguishable freshly hatched larvæ of all forms (No. 1) are very small, in no species attaining 2 mm. in length. They are delicate, feebly chitinized creatures, blind, the thoracic segments not specialized, and with short 9-to 10-jointed antennæ. After the first moult the differentiation into neuters and sexed individuals becomes appreciable, not only in the beginnings of the development of the sexual organs, but in the increase in the number of antennal joints. The larvæ and sub-

---

*By placing a small quantity of arsenic or calomel mixed with sugar in their burrows or nests, the termites will greedily devour the mixture, and by means of the poisoned individuals being fed on as fast as they perish, the whole colony will in time be destroyed.

sequent stages of the neuters remain eyeless and the thoracic segments are
very little altered, since they develop no wings. But after the second moult
a further differentiation takes place between the larvæ of the ordinary
workers and soldiers, those of the former being recognized by the small
head, smaller mandibles, large maxillæ and labium, while those of the latter have a much larger head, very prominent mandibles, variously modified
according to species, and much smaller maxillæ and labial parts. In the
perfect workers and soldiers these differences are still more strongly marked,
and both forms may at once be distinguished from other larvæ by the
darker color and the shining and harder integuments.

A peculiar form of neuter, occurring in Eutermes, the so-called nasuti,
remained a puzzle for a long time. In this form the head is pear-shaped
and prolonged anteriorly into a tube or nose which possesses a channel
leading backward into the head. The nasuti have the power of secreting a
viscid liquid from the tip of this nose. The mandibles are not prolonged
and are unfitted for biting, while the lower mouth-parts are but little better
developed than in the common soldiers. Dr. Hagen in the Appendix to
his famous monograph of the Termes, recognized this form as a soldier form,
characteristic of the genus Eutermes, which replaces the large-headed and
mandibulate soldiers of the other genera. Mr. Hubbard, however, records
having found in one colony of *Eutermes rippertii* in Jamaica a few of these
nasuti among the soldiers (Boston Soc. Nat. Hist., 1877, pp. 270–2). It is
believed, and I think justly, by Fritz Müller that when found in colonies of
other Termites having mandibulate soldiers, these nasuti are mere inquilines
or intruders, and the opposite view is justifiable, that when the mandibulate soldier is found among the nasuti, it also is an intruder.*

Acknowledgment.

Figures 1, 2, 3, 8, 9, 10 and 11, are made from illustrations belonging to
the Department of Agriculture, and are used by the kind permission of
Chas. R. Dabney, Jr., Assistant Secretary of Agriculture.

*Since this address was written, I have had an opportunity of studying Eutermes in
the West Indies, *E. morio*, at St. Thomas, St. Kitts, Monserrat, Dominica, Mar-
tinique, St. Lucia and Barbados, and both it and *E. rippertii* in Jamaica. The
nasuti are here the smallest individuals in the colony and also somewhat
the darkest. They have no power of biting, and no organ of offense, as the liquid
exuded from the tip of the nose has no pungent property. They may, therefore,
be handled with perfect impunity Of some forty nests examined none have
furnished a mandibulate soldier. The nasuti, though having no weapon of offense (so
far at least as man is concerned) are nevertheless active guards, and undoubtedly take
the place of the soldiers in Termes proper. They crowd around the queen, when the
colony is disturbed, and rush to the outside and about the borders of any breakage or
hole made in the nest or the tunnels thereto. They throw up the head and play the
antennæ and palpi in a comically threatening way, considering their inoffensiveness,
and they watch around the borders on the inside of such breakage while the workers
run up rapidly now and again to deposit the soft excrement which is to mend the gap,
and of which the tunnels and nests are for the most part formed. Eggs and young
larvæ are frequently borne on the nose and on the feelers of these nasuti; but I have not
yet satisfied myself that they are thus purposely carried, and are not accidentally stuck
by the exuding liquid, the latter view comporting best with most of the cases. But
that these nasuti perform some function in the economy of the colony other than that
of soldiery defence, is rendered almost certain by their relatively large numbers compared with the real soldiers in Termes, for they are generally as numerous as the mandibulate workers and sometimes as numerous as all the other individuals together
While the liquid from the nose may be used in cementing the walls of the tunnels, I am
inclined to believe that it is of more importance in furnishing the first pabulum of the
young.
Eutermes rippertii differs little from *E. morio* in habit except that the hard, paler
nodules generally found in its older nests do not occur in those of the latter. But the
most interesting experience, which is born out by the observations of Mr Dudley on the
species in Panama, is that I have found as many as nine queens in one nest and often
three or four. In fact there is every variation, even in independent nests which apparently have no accessory mother-nest, from those without queen to those with one up to
nine (or more according to Dudley), while in one nest I found scores of true royal pairs
in which the queens had undergone no material enlargement. I have also found either
no male or sometimes two and once three males associated with a single queen.
Ordinarily, however, there is but a pair, i. e., one queen and her escort

* 9 7 8 3 7 4 1 1 0 5 6 4 7 *